Climate Risks as Organizational Problems

This book is part of the Peter Lang Media and Communication list.
Every volume is peer reviewed and meets
the highest quality standards for content and production.

PETER LANG
New York • Bern • Berlin
Brussels • Vienna • Oxford • Warsaw

Theresa Castor

Climate Risks as Organizational Problems

Constructing Agency and Action

PETER LANG
New York • Bern • Berlin
Brussels • Vienna • Oxford • Warsaw

Library of Congress Cataloging-in-Publication Data

Names: Castor, Theresa Rose, author.
Title: Climate risks as organizational problems: constructing agency and action / Theresa Castor.
Description: New York: Peter Lang, 2018.
Includes bibliographical references and index.
Identifiers: LCCN 2017051871 | ISBN 978-1-4331-5021-0 (hardback: alk. paper)
ISBN 978-1-4331-3335-0 (paperback: alk. paper) | ISBN 978-1-4331-5093-7 (ebook pdf)
ISBN 978-1-4331-5094-4 (epub) | ISBN 978-1-4331-5095-1 (mobi)
Subjects: LCSH: Organizational sociology. | Communication in organizations.
Classification: LCC HM796 .C37 2018 | DDC 302.3/5—dc23
LC record available at https://lccn.loc.gov/2017051871
DOI 10.3726/b12609

Bibliographic information published by **Die Deutsche Nationalbibliothek**.
Die Deutsche Nationalbibliothek lists this publication in the "Deutsche Nationalbibliografie"; detailed bibliographic data are available on the Internet at http://dnb.d-nb.de/.

29 Broadway, 18th floor, New York, NY 10006
www.peterlang.com

This book is dedicated to my loving, patient,
and supportive husband, Paul, and son, Teddy.

CONTENTS

PREFACE

For the past decade, my research and way of thinking about organizations as communication has been strongly influenced by the Communication as Constitutive of Organizing (CCO) perspective. Although I have read a fair deal of excellent, cutting edge theorizing and research on the CCO perspective, it has been challenging for me to share this work and way of thinking with undergraduate students enrolled in my organizational communication classes. In part this has been due to the limited space in introductory text books dedicated to this perspective, and in part, it has been due to the challenges in explaining the complex ideas associated with the CCO perspective in ways that are easy to grasp for students who are new to organizational communication (see Kuhn & Schoeneborn, 2015).

I was first formally introduced to the Communicative Constitution of Organizations perspective as an attendee of the 2002 National Communication Association preconference, "Communication in Action: The Communicative Constitution of Organizations and Its Implications for Theory, Research, and Practice," held in New Orleans, Louisiana. The papers and presentations at the preconference subsequently formed the basis for *Building Theories of Organization: The Constitutive Role of Communication*, edited by Linda Putnam and Anne Nicotera (2008).

However, the most significant influence for me in positioning myself as a CCO scholar was through a collaborative project with CCO scholar, François Cooren. I first met François in 2004 when he was a respondent for a paper I presented at the International Communication Association conference. François' analysis regarding the data for the project I presented on were so insightful and interesting that I asked if he would be willing to work with me on a re-analysis of the data from a CCO perspective. He generously agreed to this collaboration, resulting in the publication of our project in a 2006 issue of *Management Communication Quarterly* (see Castor & Cooren, 2006).

My initial training in communication focused on social constructionism, discourse analysis, and the ethnography of communication with a primary focus on language-use in organizational decision-making interactions (see Castor, 2005). In my view, there are strong connections between discourse analysis, social constructionism, and the CCO perspective (see Bartesaghi & Castor, 2008). However, CCO and the other aforementioned areas should not be conflated given differences in many of the specific issues, concepts, and influencing theories that form the basis for CCO. A primary interest of CCO work is specifically how is the *organization* constituted and as such, what is it about communication practices specifically that contribute to the formation, maintenance, or even failure of an organization (see Bisel, 2010).

The purpose of this text is to introduce key ideas of the CCO perspective to undergraduate students. However, this work may also provide a useful introduction for graduate students and scholars who may be new to this area. After the introductory chapters, the main chapters of this work focus on a key topic for organizational communication generally, and for the CCO perspective in particular. These topical chapters feature case studies that thematize issues related to climate crises, the environment, and weather, therefore this work may also be relevant for students and scholars with an interest in environmental communication, risk communication, crisis communication, public relations, and public health. The case studies will also be of relevance for those interested in understanding the ways that governmental agencies "interact" with weather crises and the public.

Chapter one provides a general overview of the CCO perspective while chapter two explicates the three main CCO perspectives. While the latter chapter is intended to focus on the theories and concepts behind the CCO perspectives, based on student feedback on an earlier draft of this work, the theories without illustrative examples are challenging to understand.

Therefore, brief examples are provided to explain key concepts of the three main CCO perspectives.

Chapters two through six each address different key concepts in organizational communication and apply a CCO perspective to explain the key communication issues and challenges involved with each case study. These chapters are mostly based on prior research that I have conducted on a specific weather crisis or environmental issue. The analyses on Hurricane Katrina and the Great Lakes are based on work that has been published in academic journals (Castor, 2016; Castor & Bartesaghi, 2016) and the chapter on the Winter Storm was presented at the 2015 International Communication Association conference.

The concluding chapter summarizes key ideas of CCO highlighting the practical benefits of adopting such a perspective as well as potential limitations or challenges for this area to address.

References

Bartesaghi, M., & Castor, T. (2008). Social construction in communication: Revisiting the conversation. *Communication Yearbook*, *32*, 3–39.

Bisel, R. S. (2010). A communicative ontology of organization? A description, history, and critique of CCO theories for organization science. *Management Communication Quarterly*, *24*(1), 124–131.

Castor, T. (2005). Constructing social reality in organizational decision-making: Account vocabularies in a diversity discussion. *Management Communication Quarterly*, *18*, 479–508.

Castor, T. (2016). Discursively constructing the Great Lakes freshwater. In R. Browning (Ed.), *Advances in research using the C-SPAN archives* (pp. 33–57). West Lafayette, IN: Purdue University.

Castor, T., & Bartesaghi, M. (2016). Metacommunication during disaster response: 'Reporting' and the constitution of problems in Hurricane Katrina teleconferences. *Management Communication Quarterly*, *30*, 472–502.

Castor, T., & Cooren, F. (2006). Organizations as hybrid forms of life: The implications of the selection of human and non-human agents in problem-formulation. *Management Communication Quarterly*, *19*, 570–600.

Kuhn, T., & Schoeneborn, D. (2015). The pedagogy of CCO. *Management Communication Quarterly*, *29*, 295–301. doi: 10.1177/0893318915571348.

Putnam, L. L., & Nicotera, A. M. (2008). *Building theories of organization: The constitutive role of communication*. New York, NY: Routledge.

ACKNOWLEDGEMENTS

The original genesis for this work came from a conversation that I was having at my colleague, Wendy Leeds-Hurwitz' former home in Milwaukee, Wisconsin. I was seeking advice on a book project (not this one) when Wendy promptly asked me about what I envisioned for my career trajectory and where was it that I had something that I wanted to say. This pointed question was the genesis of this book project that I had been contemplating for a while in that it brings together a number of career passions for me: teaching undergraduate students, conducting and writing about research, understanding the constitutive power of communication, interrogating the complex relationship between communication, climate, and action.

I also wish to acknowledge my undergraduate organizational communication students at the University of Wisconsin-Parkside who have pushed me in bringing to the forefront the interconnection between organizational theory and practice. I also wish to acknowledge my "undergraduate research apprentices" who, with university support from the Undergraduate Research Apprenticeship Program, have worked closely with me on advanced research that goes beyond the boundaries of the classroom experience. In particular, former student Kristin Kotleski has assisted me in examining communication related to Great Lakes water policy issues that indirectly informed the chapter related to the Great

Lakes. My colleagues in my department, Fay Akindes, Linda Crafton, Jonathan Shailor, Adrienne Viramontes, and Mary Waid have provided invaluable support for excellent teaching and engaged scholarship.

There are several scholars who have provided important input on the analyses and cases described within this work. In particular, I am indebted to François Cooren for his ongoing insightful feedback and support as well as to Phillip Clampitt. I am grateful to long-time collaborator Mariaelena Bartesaghi for her support and in-depth knowledge of discourse and social construction and for her initial invitation to study the teleconferences during the Hurricane Katrina disaster. I am also appreciative of Gail Fairhurst, aside from her excellent scholarship, there have been several times in which she has graciously agreed to read drafts of my papers related to CCO. While these and other scholars have been important influences on my writing, I accept full responsibility for errors and shortcomings within this work!

I am very grateful to Kathryn Harrison and the staff of Peter Lang for their support and encouragement of this project.

This work is for the next generation as our world faces challenging problems with climate change. As a society, our first line of action is to determine as social collectives, how we wish to act and what future we wish to constitute.

Finally, I wish to express my appreciation to my family, husband Paul Hopton and son, Teddy Hopton, who have tolerated many hours of their wife and mother's time in front of the laptop!

PART 1

THEORETICAL AND CONCEPTUAL BACKGROUND

· 1 ·

INTRODUCTION

> "It was an organizational decision."
> "We aren't issuing licenses today."

Every day, people speak in terms of *organizations* where the organization seems to be an entity that can dictate what people can or cannot do. The above statements illustrate how people use language to construct organizations. In the first example, taken from a press conference in 2013, Green Bay Packers coach Mike McCarthy was explaining the "decision" for keeping star quarterback Aaron Rodgers out of a seventh straight game due to a collarbone injury. In this statement, responsibility for the decision is given to an "organization" rather than to the individuals who participated in the decision-making process. The second example is taken from a statement uttered by a Kentucky County Clerk, Kim Davis, in her response to a same sex couple that had applied for a marriage license. In the second example, by referring to "we" rather than "I" when pointing to who is responsible for a decision, the organization rather than the individual is given responsibility for the action. These examples get at the idea that organizations are *constituted* through communication. This is also the key premise of what is known as the "Communication as Constitutive of Organizations" or CCO perspective.

Organizational communication theories, like all theories, evolve and change as new thinking and research emerges on a given topic. Within the field of Organizational Communication, one of the newest general perspectives is the CCO approach. While there is a great deal of theorizing and research in this newer area of organizational communication, that work has been written primarily to address other scholars and researchers. However, as Kuhn and Schoeneborn (2015) noted, one of the challenges in introducing the CCO perspective to students is that it has complex ideas explained in complex ways.

The CCO approach generally views organizations as being constituted through communication. It stems from the interpretive and discursive turns in organizational studies (see Fairhurst & Putnam, 2004; Putnam & Pacanowsky, 1983). A key claim of the CCO approach deals with the ontological nature of organizations: communication does not merely reflect an organizational reality, but rather organizations are constructed through communication. While this general assumption is shared by CCO scholars, there are differences among CCO approaches (Bisel, 2009). CCO approaches may be divided into three approaches: the Montreal School (TMS) approach, the Four Flows model, and the Social Systems (TSS) approach (Schoeneborn et al., 2014).

The purpose of this work is to provide an introduction to the Communication as Constitutive of Organizations approach and to illustrate ways to apply it for understanding how organizations are constituted or constructed through communication. This work will address key ideas of organizational communication such as sensemaking, decision-making, problem-formulation, and agency, from a CCO perspective. A case study approach is taken where the focus of each case is on a climate-related issue such as a weather-related crisis or natural resource.

Weather, climate, and environment-related case studies were selected given their importance in and of their own right. Each year worldwide, hundreds of millions of people are affected by natural hazards at an economic cost of hundreds of millions of dollars per year. During the time span of 2004–2014, the United Nations Office for Disaster and Risk Reduction estimates that 1.7 billion people were affected by natural hazards that resulted in 700,000 deaths and cost $1.4 trillion in damage. In addition, crises such as the Flint, Michigan lead water contamination problem highlight the interdependence between society and natural resources and the importance of organizational management of our natural resources and of communicating responsibly.

While natural hazards are worthy of study generally because of their impact, they are also worthy of study from an organizational communication perspective. Organizations such as governmental agencies, international organizations, nonprofit organizations, and nongovernmental organizations (NGO), among others have had some role in preparing for or helping people to recover from natural hazards. Meteorologists with weather service agencies or media stations make weather forecasts. Governmental agencies may be responsible for preparing people for potential natural hazards (e.g., severe cold weather preparations, heat wave warnings). Governmental officials make decisions on how to prepare for forthcoming hazards (e.g., should the schools be closed for the winter storm? should the city be evacuated for the hurricane?). In the aftermath of a crisis or disaster, organizations also play a vital role in recovery for a community (e.g., finding and assisting survivors). Given that natural hazards are ongoing yet have a degree of unpredictability, examining how organizations respond to natural hazards provides a fitting circumstance for studying constitutive processes that are centered on key organizational activities such as sensemaking, decision-making, and problem-solving.

In the remaining sections of this chapter, I provide a general introduction to the Communication as Constitutive of Organizations approach, first describing its historical development and then explaining key ideas of the CCO perspective. The chapter concludes with a preview of the rest of the book.

Communication as Constitutive of Organizations

The CCO perspective is not a specific theory, but a general perspective related to assumptions about what is an organization and what is the relationship between communication and organizing. CCO scholars argue that organizations are constituted or constructed through communication—without communication, there is not an organization. There are specific CCO perspectives that share this general assumption, but they describe and emphasize different processes, and in certain ways, are in tension with each other.

Conceptual and Theoretical Roots

While there are early roots and precursors, Ashcraft, Kuhn, and Cooren (2009) traced the initial use of the phrase "communicative constitution of organizations" to Putnam and Nicotera's 2009 edited book, *Building Theories of Organization: The Constitutive Role of Communication.* This book was developed from a 2002 National Communication Association preconference sponsored by the Organizational Communication Division titled, "The Communicative Constitution of Organization and Its Implications for Theory, Research, and Practice." The CCO approach is part of an evolution in thinking about organizations that can be traced back to interpretive approaches to organization.

From Functional to Interpretive Perspectives in Organizational Communication

The functional perspective in organizational communication refers to a perspective in which organizations and their components are viewed as objective facts (Putnam, 1983). In contrast to viewing organizations as being constituted through communication, this view treats organizations as metaphorical "containers" in which communication occurs. In other words, the organization exists independent of communication, and when communication occurs, the organization is just a place or container for that communication. In this view, communication is treated as "messages" that can flow up and down a chain of command. Messages may consist of information or commands and orders. The importance of this latter statement can be further explained by looking into the "information transmission" model of communication.

One of the first theories of communication is known as the *information transmission model* of communication derived from the work of Shannon and Weaver (1949). These scholars were not communication scholars per se; rather, they were engineers addressing issues with telephone technology. Thus, they literally were concerned with the problem of how to transmit information across distances using communication technology (e.g., telephones, radio).

Although Shannon and Weaver were addressing the specific issue of communication technology, their general idea of "communication as information transmission" is one that has greatly influenced the development of communication theory. This view can be summarized as one that assumes that communication is a process in which a sender sends a message to a receiver through a channel. Communication is used as a tool and the substance of communication is information. A simple comparison is with the childhood game of "telephone" in which one child may whisper a phrase into the ear of another child who then tries to repeat the same phrase to the next child in a line. If the last child has received the message as the first child stated, then communication can be said to have occurred successfully. The information transmission model is thus concerned with the accuracy of how messages are transmitted and factors that may inhibit successful transmission such as background noise.

A key implication of the information transmission model is that the main purpose of communication is to *represent* what is in the head of the sender. If the receiver understands the message as conceptualized by the sender, then communication has been successful. A key limitation of this perspective is that communication serves multiple purposes beyond representing: it can have multiple and ambiguous meanings, it can develop relationships. These nuances of communication are not captured in the information transmission model.

The information transmission model is reflected in early theories of organizational behavior such as *Frederick Taylor's theory of scientific management* (see Eisenberg, Trethewey, LeGreco, & Goodall, 2017; Mumby, 2012). The key premise of Frederick Taylor's approach is that activities in organizations (specifically, factories) can be analyzed in a scientific manner to determine how to work most efficiently. Taylor's approach involved breaking down the steps of an activity and trying to determine how to minimize the number of steps needed to complete a task. The main purpose of communication is functional in that it is used to convey commands and information from managers to workers.

In contrast to functional perspectives, *interpretive approaches* view communication as symbolic and as creating meaning rather than conveying information (Putnam & Pacanowsky, 1983). These approaches emphasize that communication is an interpretive process and that the goals and outcomes of communication involve more than conveying information, but also creating meaning and relationships. Interpretive approaches differed by focusing on communication in terms of symbols and meanings rather than information. To understand meaning, you must also understand social relationships, identities, culture, context, etc. In addition, meaning is not static, but fluid, dynamic, changing and multifaceted.

A classic example of a study that takes an interpretive perspective is Smith and Eisenberg's (1987) study of conflict at Disneyland for which they analyzed "root metaphors" that described the conflict. The presence of metaphors in organizational communication right away challenges the application of an information transmission view to organizational communication. A metaphor is a figure of speech that describes one thing in terms of something else that is dissimilar to it (e.g., Shakespeare's phrase that "all the world's a stage"). The communication significance of metaphors relies on creating meaning through new comparisons.

Disneyland has been called the "happiest place on earth," but Smith and Eisenberg's study of a major organizational conflict in the early 1980's revealed otherwise. First, communication at Disneyland itself is filled with metaphors as employees speak of being "cast members" (rather than workers) and refer to their "costumes" (rather than uniforms) and "roles" (rather than jobs) and the "show" (rather than work). An especially strong metaphor that Smith and Eisenberg identified is that of "family," however, the idea of "family" was in conflict with managerial decisions that involved laying-off individuals. As an interpretive study, Smith and Eisenberg's work illustrates how the meanings of communication as symbolic are important for understanding organizational communication dynamics.

The interpretive perspective is closely aligned with a constitutive perspective in that both describe communication as a subjective process of meaning making. However, the constitutive approach differs in that a key issue it addresses deals with the *substance* of what makes up an organization: communication *constitutes* organizations. What this also implies is that organizations are not "things" (such as containers). Rather, organizations are processes—a key point that organizational theorist Karl Weick helped to introduce.

Organization/Organizing: The Importance of Nouns and Verbs

Organizational theorist Karl Weick has made several important contributions that laid the groundwork for a constitutive perspective of organizations. A simple yet important point that he made was in arguing that organizations were not nouns, but rather, consisted of verbs (Weick, 1979). This difference is significant in that nouns are things, entities or beings whereas verbs are about actions and processes. Weick argued that we need to conceptualize of organizations as consisting of processes rather than as static, pre-existing entities: instead of organization, we need to consider organizing in that without these processes, what we know as an organization would not be possible.

Weick's revelation points to the importance of understanding the processes of organizing. This is an issue that he develops in his *sensemaking* approach (Weick, 1979, 1995). Briefly defined, sensemaking is the process of giving meaning to experience. This concept points out that humans experience and construct the world based on their subjective filters. The concept of sensemaking itself is a multidisciplinary concept, but was popularized in organization studies through Weick's work.

Because of the subjective nature of how people make sense of their experiences, how we then act is based on what we believe is occurring. Weick used the idea of *enactment* to highlight this in that organizational environments are constructed rather than pre-made. Weick's work on crisis helps to explain these ideas as well as emphasize why it is important to understand organizational crises through a sensemaking lens.

A classic example from Weick's work that illustrates some of the aforementioned ideas is his study of the *Mann Gulch fire disaster* (Weick, 1993). This disaster occurred in 1949 in Montana when a lightning storm is believed to have set fire to a dead tree in turn causing a forest fire. To deal with the fire, a crew of 16 smokejumpers were flown in from Missoula, Montana. Due to a series of tragic steps and circumstances, the lives of 13 of the smokejumpers were lost in the forest fire. The forest fire was a more challenging one to deal with—it was an extremely hot day (97°F) with unexpected high winds. This resulted in a fire of unusually hot temperatures where its course through the forest was difficult to predict. The Mann Gulch disaster is the topic of a book by Norman Maclean (1993) entitled, *Young Men and Fire*. The story of the fire disaster as re-told in Maclean's book is the focus of study for Weick.

In his analysis, Weick (1993) explained the Mann Gulch fire disaster as due in part to a failure in sensemaking. On August 5, 1949, the day of the fire, there were several events that occurred that contributed to the smokejumpers inability to handle the fire and to save themselves:

1. On the morning of the forest fire, when it was spotted, it was referred to as a "10:00 fire," which, in smokejumper lingo, implies that it can be dealt with quickly.
2. Two key smokejumper leaders eat dinner on the hike toward the fire—the rest of the crew wonder how the fire can be that serious if the leaders feel comfortable enough to eat dinner?
3. One of the leaders runs away rather than toward the river at one point; this same leader also tells the other smokejumpers to "drop your tools." These actions on the part of a leader did not make sense to the other smokejumpers.

The above list is partial (Weick identified additional factors). The key point that Weick makes in his analysis here is that there is a close interconnection between meaning and identity in how the crew handled and reacted to the fire. There were terms, phrases, and actions that implied the fire was not as severe as it was; the command to "drop your tools" and other actions went against what the smokejumpers had been taught to believe about themselves and how to fight a forest fire. However, if they would have dropped their tools, they could have ran more quickly.

There were three smokejumpers who survived the fire. How did they do this? One of the survivors, the group leader, "Wag" Dodge created an "escape fire" where he developed a small, controlled fire that would take away fuel from a larger, approaching fire. The creation of an escape fire is now a common practice for escaping a wildfire thanks in part to Dodge's innovative use of it during the Mann Gulch disaster. The other two survivors, Robert Sulley and Walter Rumsey, unlike most of the rest of the crew, stuck together and were able to find a rocky area where the fire bypassed them.

What does the Mann Gulch disaster illustrate regarding sensemaking? Making sense of the fire itself was a subjective process where labels and not just smoke and flames participated in constituting the fire as an emergency. Ways of dealing with the fire intersected with the identities and prior assumptions of the smokejumpers where orders that involved running away from a river, setting a little fire, and "dropping your tools" did not make sense with what they understood about how to fight let alone survive a wildfire. Communication constituted their understanding of how to organize as smokejumpers to fight a wildfire.

The CCO approach and the concept of organizational sensemaking challenge and re-frame what organizations are and how they should be conceptualized in terms of processes. However, some organizational communication scholars have found aspects of Weick's work unsatisfactory. For example, while he describes sensemaking as a social and interpretive process, he does not explain exactly how this occurs (see Castor & Cooren, 2006; Cooren, 2001). Nonetheless, Weick's sensemaking approach and insight regarding *organizing* (rather than organization) has provided an important basis for the CCO approach. In a later chapter, Weick's sensemaking approach will be re-visited and expanded on.

Communication as "Constitutive"

"Constituting" or "constitutive" gets at the idea that organizations are created and made up of communication. The Merriam-Webster online dictionary defines "constitute" as: "to make up or form something; to be the same as something: to be equivalent to something; to establish or create (an organization, a government, etc.)" As McPhee and Zaug (2000) explain of constitution:

> "Constitution" has a variety of philosophical meanings ranging between the epistemological (we, as researchers, know reality by constituting concepts of it) and the ontological (we, as members of society, constitute our reality, or social reality), between the cognitive and the practical/active, between creation and making sense of what already exists. (Outhwaite, 1983, pp. 26–27)

The evolution of theorizing in organizational communication has paralleled that of the broader discipline of communication. For example, Deetz (1994) compared and contrasted views of communication as "expressive" versus "constitutive". The expressive view is aligned with the information transmission view described earlier. Deetz explained that:

> The expression conception draws attention to interaction primarily in reproductive terms—expression enables reproduction of meanings, perceptions, or feelings that reside independently somewhere else. In the *constitutive* conception, messages are an active part of the production of meaning, perceptions, and feelings. (pp. 572–573)

The difference between the two perspectives is that the expressive view assumes that communication is about something whereas in the constitutive view, communication functions to construct the very things we are talking about.

Carey (2008) contrasted a "transmission" view of communication with a "ritual" perspective. The ritual perspective describes communication as part of human social rituals where the importance of communication isn't so much that we are giving and receiving information (the transmission view), but rather, that we are using communication to be together in community. For example, consider how, when you get together with friends and/or loved ones, you may joke around. That joking is a part of your group ritual to reinforce your sense of community and social connection. As another example, consider the act of watching a sporting event on television. The necessary information on what is going on in the game is conveyed by the announcer on the television network, yet, sports fans will still "communicate" through cheers, yells, and their own commentaries on the game.

Stewart (1995) makes the point that our use of language is interconnected with our sense of who we are as humans. Moreover, it is not just the use of language by individuals that makes us human, but rather, it is our use of language in interactions (i.e., "articulate contact"). To use a metaphor that John Stewart regularly used when I attended his graduate seminars, humans without communication would be like taking a fish out of water: our sense of the world would radically change so much so, that our sense of identity and even our ability to survive would be threatened.

The constitutive perspective in communication is a view that was advanced a few decades ago and as such, has influenced the diverse discipline of communication in many ways. The constitutive perspective in the broader communication discipline has helped in opening up disciplinary, intellectual space for thinking of communication as constitutive of organizations.

The Social Construction of Reality

Another parallel yet related development in theorizing that provided support for understanding the communicative constitution of organizations is the social constructionist perspective. The phrase "social construction of reality" was popularized through Berger and Luckmann (1966). The premise of a social constructionist perspective is that what we take to be real is constructed through our interactions. In other words, what we know as reality is something that we do not objectively know. Rather, what we know about the world is subjectively-based on many factors that include individual perceptions, culture, and other social influences.

Putnam, Nicotera, and McPhee (2009) noted the similarities between the CCO perspective and a social constructionist perspective in that "both focus on forming, making, and composing" (p. 4). However, they also noted differences of emphasis between the two perspectives. A CCO perspective "tends to 'unmask' a phenomenon" and emphasizes "its historical emergence and contemporary reproduction, practices of organizing and participating, a body of knowledge about organizing, and the organization as a reified thing" (p. 4).

While the CCO approach shares in common with a social constructionist approach the belief that reality is socially constructed, they differ in some of the specific issues and features that a CCO approach raises. First, the central question for the CCO perspective is how are organizations communicatively constituted? Second, the CCO perspective prioritizes addressing the role of communication. The social constructionist perspective is so broad and multidiscplinary with strands in diverse areas such as psychology, sociology, education, geography, and environmental studies as well as communication, that the extent to which communication processes are addressed typically depends on the disciplinary background of the researcher (see Bartesaghi & Castor, 2008). Even though the CCO perspective has developed in ways that are distinct from a social constructionist perspective, they both share in common a belief that reality is socially and subjectively created and maintained through human interaction (Galanes & Leeds-Hurwitz, 2009).

Summary

This review of conceptual and theoretical roots was intended to emphasize some of the key assumptions of the communication as constitutive of organizations perspective. The CCO perspective is part of an ongoing evolution in organizational communication theorizing that moved from viewing communication as functional to interpretive. This shift was ushered in partly due to the contributions of Karl Weick's sensemaking perspective that highlighted how organizations should be conceptualized in terms of "verbs" (i.e., processes) rather than "nouns" (i.e., entities, containers). The development of the CCO perspective is also part of a larger, broader communication disciplinary perspective that shifted from an information transmission view to a constitutive view of communication. This shift is important in emphasizing the significance of communication in building and sustaining community and in constructing social reality.

Key CCO Concepts

Within the CCO perspective, there are several important key phrases and concepts. Some of these concepts have been explored in the preceding section and all of these concepts will be extended upon in the analysis chapters. This section is intended to provide a brief overview of the following key CCO ideas: communication, constitution, organization/organizing, discourse, agency, and materiality.

Communication

There are several definitions of communication generally; the view of communication that CCO approaches assume is that communication is a constitutive process (Putnam et al., 2009). As such, it is ongoing and can include symbols, meanings, and language. It also includes nonverbal communication, objects, and places. It comes in many forms that include speaking, writing, and technologically-mediated communication. While communication is a basis of CCO approaches, there are differences across specific CCO perspectives in how communication is characterized. These differences are addressed in the following section.

Constitution/Constituting

After exploring the nuances of the term "constitution" from everyday dictionaries, Putnam et al. (2009) noted:

> The typical definition of "constitute" and "constitution" highlights the forming, composing, or making of something in addition to describing the phenomenon that is

constituted. These definitions also emphasize the mode or state in which something is formed. (pp. 3–4)

Constitution emphasizes how something is constructed or created. However, CCO approaches go beyond the basic claim that reality is socially constructed and include other features of situations as part of understanding how organizations are communicatively constituted. For instance, some CCO approaches include may critically analyzing ("unmasking") the exercise of power or examine the relationship between history and the development of an organization's practices.

Organization/Organizing

CCO approaches view organizations as processes rather than products, "containers," or "entities." Because this distinction has been explored already in this chapter, here I focus on the basic idea of what many scholars mean when they say they are studying organizational communication. In other words, what makes something organizational? While CCO scholars focus on organizations as processes, there are certain types of social groupings that have been studied as organizations.

Charles Redding, who is largely considered the founder of the field of organizational communication, explained that complex organizations have the following features:

(1) interdependence among organizational members;
(2) specific tasks and functions assigned to individuals and units;
(3) goals to achieve as a collective; and,
(4) the exercise of control over the individuals within the organization (Redding, 1988).

Within organizations, as compared to other social groupings, there is a degree of formalization. Typically, there is some sort of legal formalization or sanctioning of the organization (e.g., a nonprofit organization must be recognized as such through an application and federal sanctioning).

Organizations may include business, corporations, firms, nonprofit organizations, governmental agencies, cooperatives, etc. So, while a family is a type of grouping that can be said to be organized, the study of families has typically been relegated to the field of family communication or interpersonal communication. While many of the characteristics identified by Redding could be applied to families, they are not what define a family. For example, if a child stops doing his chores, he will not be fired from the family. However, a family may start a family business if the members decide to market and sell a product and complete other paperwork to be recognized formally as a business.

Discourse

While not necessarily common across all CCO perspectives, *discourse* is an important concept and area of research for many CCO scholars. "Discourse" has several, inter-related meanings (see Alvesson & Karreman, 2000). It can refer to the specific words and language that people use, or what is sometimes referred to as the "little-d" discourse. It can also refer to an interrelated set of ideas and ways of thinking about a topic, also known as "big-D" Discourse. Putnam et al. (2009) described discourse as "a sub-set of communication, one that excludes networks, information processing, or non-linguistic forms of organizing" (p. 8). For some CCO scholars, particularly those within what is known as the Montreal School approach, the examination of naturally-occurring language practices is an important aspect of understanding how organizations are communicatively constituted.

In general, discourse has become an important aspect of thinking about organizational communication. Fairhurst and Putnam (2004) argued that organizations are "discursive constructions" and identified three "orientations" to the relationship between discourse and organizations: an object orientation, the "becoming" orientation, and the "grounded in action" orientation. The object orientation is also known as the container approach and focuses on discourse as produced by organizations. The becoming orientation examines how discourse creates organizations and emphasizes organizations as social constructions that are created and sustained through communication. The grounded in action orientation is similar to the becoming orientation but emphasizes communication as occurring over time by acknowledging the communication that occurred before and will occur after a given point in time. This approach also examines how humans and objects relate to each other in constituting organizations. This latter approach is also well-aligned with particular CCO perspectives.

Putnam et al. (2009) identified the following as intellectual roots of the CCO perspective: speech act theory (Austin, 1962; Searle, 1969), rules theory (Cushman, 1977), systems theory (Luhmann, 1995), ethnomethodology (Garfinkel, 1967), conversation analysis (Sacks, Schegloff, & Jefferson, 1974), frame analysis (Goffman, 1959), critical discourse analysis (Fairclough, 1995), structuration (Giddens, 1984), semiology (Barthes, 1954/1967), narrative theory (Greimas, 1987), and critical theory (Derrida, 1988; Foucault, 1972; Heidegger, 1959). Many of these deal directly or indirectly with language use. For example, speech act theory, conversation analysis, frame analysis, and narrative analysis have been applied in examining language-use in social interactions for a little "d" discourse approach. Approaches such as critical theory have been used to examine big "D" discourses related to organizational communication.

Agency

A key issue for CCO scholars is the notion of *agency* in the sense of who (or what) has the capacity to act. Because organizing is an activity, agency is intertwined with organizing when considering how organizational actions are accomplished whether it is about decision-making, policy implementation, or strategic planning (to name a few). As such, agency is a key concept that CCO scholars have discussed. It is also an area of controversy among CCO scholars given the divergent ways that different CCO scholars have conceptualized of agency (a topic that is discussed in greater detail in other chapters).

Materiality

Closely connected to the disagreement about agency, the CCO approaches also have different ideas they emphasize regarding *materiality*. Leonardi (2012) explained that materiality consists of "matter" and "form." For organizations, materiality may include the tools and technologies of the organization such as desks, copying machines, hallways, post-it notes, computers, widgets, etc.

However, Leonardi cautions that materiality should not be confused with "physicality" in that we may also speak of "digital" materials. A key point to emphasize about materiality is that CCO approaches take into consideration how the material, physical, or sociomaterial is a part of how organizations get constituted. Moreover, these materials are not just a part of the background equipment nor do we just make up what they mean. Rather, there are specific aspects of their physical features that are important for organizational communication. As a basic example, consider the disruption to a presentation that occurs when the PowerPoint does not work correctly. While the speaker could convey the information without the PowerPoint, the technology break down creates a change in the flow of interaction. While materiality is an important aspect of organizational communication, there are different ways that CCO approaches have viewed materiality.

Communicative Constitution in the Age of Climate Change

Communication makes a difference. As Penman (1992) writes:

> If there are elements of our condition that are kept rationally invisible or are only incipient in our practices, then our communication has an impoverished representation of what could be possible. This in turn limits our options and opportunities. (p. 241)

A powerful implication of the constitutive perspective is that it calls attention to the *creating* or *constructive* aspect of communication to shape our social *and* physical worlds. In turn, it also emphasizes how our social and physical worlds may shape us. With awareness and understanding, rather than having our (organizational) conditions be "kept rationally invisible," individuals may be empowered in their choices about their organizational communication practices so that we may expand, rather than limit, our "options and opportunities."

The constitutive approach and a goal of this specific book is to get beyond basic binaries that have been involved in discussions of social constructionism. One critique of social constructionism is that it assumes that everything is relative, that reality can only be socially constructed by people, and that all points of view are equally valid. In addressing issues such as materiality and non-human agency, CCO approaches help to show how both people and non-human entities and phenomena (such as a weather event or natural hazard) can participate in constituting a crisis. Therefore, what is at issue with weather-events is not whether or not they occur, but rather, how we respond to them. This line of reasoning can be applied to climate change. There is enough scientific evidence and consensus in the scientific community to support that climate change is happening. However, what we constitute through our communication is how we organize to respond to those changes.

This project is intended to contribute to theorizing on how organizations are communicatively constituted. In expanding upon how to apply a CCO perspective, the subsequent chapters address weather-related crises and how organizations have responded to certain crises. The choice to address weather-related crises is strategic in highlighting conceptualizations of agency and materiality in the process of problem-construction. The existence of weather is undeniable and its effects are undeniable. However, this status allows for us to focus on the communication processes involved with constructing extreme weather phenomena specifically *as a crisis event*. In this respect, this process also provides a critical examination of the relationship between materiality and the communicative constitution of organizations. Events such as hurricanes, drought, fire, or cold cannot be ignored or "talked away." However, their ontological status as problem, disaster, crisis, etc. *is* something that is communicatively constituted.

Outline for Remainder of the Book

The next chapter provides an explication of three key CCO perspectives: the Montreal School (TMS) approach, the Four Flows model, and the Social Systems (TSS) perspective. While in one conceptualization of this book, I would have then liked to have applied each of those perspectives to analyze my case studies, I did not feel that I had the room or expertise to take on such an undertaking in a way that would have done each of the perspectives justice. My own studies and background in the analysis of language-use favor the Montreal school approach. Therefore, the key CCO perspective that I draw upon for the analyses in chapter three through six is influenced by this approach.

The extended case studies that are discussed in Chapters three through six of this work provide examples that powerfully illustrate why it is important to understand how communication constitutes organizations and crises. Each of these chapters highlights a specific, significant concept in organizational communication and uses that concept to frame the analysis of a natural hazard related case study. Chapter three examines how crises are communicatively constituted and the role of pace in how situations are identified as crises. This chapter takes a generalized approach to the case study focus by examining some of the challenges involved with defining and preparing for heat waves in Europe. Chapter four explains the concept of organizational sensemaking to examine the results of delayed sensemaking in preparing for a winter storm. The case study featured in this chapter is when a severe winter storm hit the southeast U.S. The concept of decision-making is also emphasized in this chapter. Chapter five expands on the issue of agency and the role that the different ways agencies may be constituted can influence the way that emergency managers and others respond to crises. This chapter focuses on the Hurricane Katrina disaster. Chapter six takes up the issue of sensemaking again, but examines how *prospective* sensemaking or anticipating a future problem led to the development of a freshwater policy to prevent a possible future crisis. The case study of focus in this chapter is the congressional discussion leading to the approval of the historic Great Lakes Compact and the key CCO concept of focus is the idea of actions nets. Chapter seven concludes this work by providing a summary and by re-visiting the utility of applying a CCO perspective for key organizational communication activities and concepts such as decision-making, sensemaking, crisis management, agency, and materiality.

The case study chapters are organized around the following themes that address different, critical questions with regards to the communicative constitution of organization and crises:

- What is going on?
- What are the possible consequences?
- Who can do something about it?
- What can be done?

While all of the above questions could be applied to the cases featured in each chapter, there are important, distinctive features of the cases that help to highlight a particular question from the above list.

If we can understand how our communication practices relate to crises and disasters and can even constitute or create those crises, then perhaps we can exercise agency in our communication choices to avert those disasters. The CCO approach helps to show how, while material circumstances such as the weather, may seem to be out of the control of human hands, how we cope with such circumstances are interconnected with the processes of organizing and communicating.

References

Alvesson, M., & Karreman, D. (2000). Varieties of discourse: On the study of organizations through discourse analysis. *Human Relations*, 53(9), 1125–1149.

Ashcraft, K. L., Kuhn, T. R., & Cooren, F. (2009). Constitutional amendments: "Materializing" organizational communication. *The Academy of Management Annals*, 3(1), 1–64. http://doi.org/10.1080/19416520903047186

Austin, J. L. (1962). *How to do things with words*. Oxford: Oxford University Press.

Bartesaghi, M., & Castor, T. (2008). Social construction in communication: Re-constituting the conversation. *Communication Yearbook*, 32, 1–39.

Barthes, R. (1954/1967). *Elements of semiology* (A. Lavers & C. Smith, Trans.). Boston, MA: Beacon Press.

Berger, P. L., & Luckmann, T. (1966). *The social construction of reality. A treatise in the sociology of knowledge*. New York, NY: Irvington Publishers.

Bisel, R. S. (2009). On a growing dualism in organizational discourse research. *Management Communication Quarterly, 22*, 614–638.

Carey, J. W. (2008). *Communication as culture, revised edition: Essays on media and society*. New York, NY: Routledge.

Castor, T. R., & Cooren, F. (2006). Organizations as hybrid forms of life. *Management Communication Quarterly, 19*(4), 570–600.

Cooren, F. (2001). Translation and articulation in the organization of coalitions: The great whale river case. *Communication Theory, 11*(2), 178–200. https://doi.org/10.1111/j.1468-2885.2001.tb00238.x

Cushman, D. (1977). The rules perspective as a theoretical basis for the study of human communication. *Communication Quarterly, 25*, 30–45.

Deetz, S. (1994). Future of the discipline: The challenges, the research, and the social contribution. *Communication Yearbook, 17*, 565–600.

Derrida, J. (1988). *Limited inc*. Evanston, IL: Northwestern University.

Eisenberg, E. M., Trethewey, A., LeGreco, M., & Goodall H. L. Jr. (2017). *Organizational communication: Balancing creativity and constraint* (8th ed.). Boston, MA: Macmillan Higher Education.

Fairclough, N. (1995). *Critical discourse analysis: The critical study of language*. London: Longman.

Fairhurst, G. T., & Putnam, L. L. (2004). Organizations as discursive constructions. *Communication Theory, 14*(1), 5–26.

Foucault, M. (1972). *The archeology of knowledge* (A. M. Sheridan Smith, Trans.). New York, NY: Pantheon. (Original work published in 1969).

Galanes, G. J., & Leeds-Hurwitz, W. (Eds.). (2009). *Socially constructing communication*. Creskill, NJ: Hampton.

Garfinkel, H. (1967). *Studies in ethnomethodology*. Englewood Cliffs, NJ: Prentice-Hall.

Giddens, A. (1984). *The constitution of society: Outline of the theory of structuration*. Cambridge: Polity.

Goffman, E. (1959). *The presentation of self in everyday life*. New York, NY: Doubleday Anchor.

Greimas, A. (1987). *On meaning: Selected writings in semiotic theory* (P. J. Perron & F. J. Collins, Trans.). Minneapolis, MN: University of Minnesota.

Heidegger, M. (1959). *An introduction to metaphysics* (R. Manheim, Trans.). New Haven, CT: Yale University.

Kuhn, T., & Schoeneborn, D. (2015). The pedagogy of CCO. *Management Communication Quarterly, 29*, 295–301. http://doi.org/10.1177/0893318915571348

Leonardi, P. M. (2012). Materiality, sociomateriality, and socio-technical systems: What do these terms mean? How are they different? Do we need them? In P. M. Leonardi, B. A. Nardi, & J. Kallinikos (Eds.), *Materiality and organizing: Social interaction in a technological world* (pp. 25–48). Oxford: Oxford University.

Luhmann, N. (1995). *Social systems* (J. Bednarz Jr. & D. Baecker, Trans.). Stanford, CA: Stanford University.

Maclean, N. (1993). *Young men and fire*. Chicago, IL: University of Chicago Press.
McPhee, R. D., & Zaug, P. (2000). The communicative constitution of organizations: A framework for explanation. *Electronic Journal of Communication*, *10*(1/2), www.cios.org/getfile/McPhee_V10n1200.
Merriam-Webster Dictionary. (n.d.). Constitute. Retrieved from http://www.merriam-webster.com/dictionary/constitute.
Mumby, D. K. (2012). *Organizational communication: A critical approach*. Thousand Oaks, CA: Sage.
Outhwaite, W. (1983). *Concept formation in social science*. London: Routledge and Kegan Paul.
Penman, R. (1992). Good theory and good practice: An argument in progress. *Communication Theory*, *2*, 234–250.
Putnam, L. L. (1983). The interpretive perspective: An alternative to functionalism. In L. L. Putnam & M. E. Pacanowsky (Eds.), *Communication and organizations: An interpretive approach* (pp. 31–54). Beverly Hills, CA: Sage.
Putnam, L. L., & Nicotera, A. M. (Eds.). (2009). *Building theories of organization: The constitutive role of communication*. Madison, NY: Routledge.
Putnam, L. L., Nicotera, A. M., & McPhee, R. D. (2009). Introduction. In L. L. Putnam & A. M. Nicotera (Eds.), *Building theories of organization: The constitutive role of communication* (pp. 1–19). Madison, NY: Routledge.
Putnam, L. L., & Pacanowsky, M. E. (1983). *Communication and organizations: An interpretive approach*. Beverly Hills, CA: Sage.
Redding, W. C. (1988). Organizational communication. In E. Barnouw (Ed.), *International encyclopedia of communications* (Vol. 3, pp. 236–239). New York, NY: Oxford.
Sacks, H., Schegloff, E. A., & Jefferson, G. (1974). A simplest systematics for the organization of turn-taking for conversation. *Language*, *50*, 696–735.
Schoeneborn, D., Blaschke, S., Cooren, F., McPhee, R. D., Seidl, D., & Taylor, J. R. (2014). The three schools of CCO thinking: Interactive dialogue and systematic comparison. *Management Communication Quarterly*, *28*(2), 285–316. https://doi.org/10.1177/0893318914527000
Searle, J. R. (1969). *Speech acts*. London: Cambridge University.
Shannon, C. E., & Weaver, W. (1949). *The mathematical theory of communication*. Cambridge, MA: Michigan Institute of Technology.
Smith, R. C., & Eisenberg, E. M. (1987). Conflict at Disneyland: A root metaphor analysis. *Communication Monographs*, *54*(4), 367–380.
Stewart, J. R. (1995). *Language as articulate contact: Toward a post-semiotic philosophy of communication*. Albany, NY: State University of New York.
United Nations Office for Disaster and Risk Reduction. (n.d.). The economic and human impact of disasters in the last 10 years [infographic]. Retrieved from https://www.flickr.com/photos/isdr/16111599814/.
Weick, K. (1979). *The social psychology of organizing*. New York, NY: Random House.
Weick, K. E. (1993). The collapse of sensemaking in organizations: The Mann Gulch disaster. *Administrative Science Quarterly*, *38*, 628–652.
Weick, K. E. (1995). *Sensemaking in organizations*. Thousand Oaks, CA: Sage.

· 2 ·

A TALE OF THREE PERSPECTIVES

McPhee (1988) described the Communicative Constitution of Organizations perspective as "meta-theory":

> guiding theorization and methodology without, typically, constituting an explanatory theory. Thus, it could guide inquiry into multiple theories about, for instance, the importance of narrative and identification in CCO thinking, or about the constitution of formal hierarchies. (McPhee, 1988) (see Schoeneborn et al., 2014, p. 289)

Brummans, Cooren, Robichaud, and Taylor (2014) summarize the general CCO perspective as having a distinctive approach to "theorizing and analyzing how organizations as discursive-material configurations are reproduced and coproduced through ongoing interactions" (p. 173). The aforementioned phrasing highlights key issues that CCO perspectives address such as the roles of discourse and materiality in constituting organizations.

There are three main CCO perspectives: the Montreal School (TMS) approach, the Four-Flows model, and the Social Systems (TSS) approach (Brummans et al., 2014). The first approach is based on the work of several scholars from the University of Montreal who have been influenced by various, mostly French, theorists such as Bruno Latour, Algirdas Greimas, and Jacques Derrida, among others. The Four-Flows model draws from British sociologist Anthony Giddens' Structuration Theory. The Social Systems approach is based on the work of German sociologist Niklas Luhmann.

As referenced to by McPhee, the different approaches illustrate varied ways to think about the relationship between organizing and communication. Each is influenced by different scholars and background theories. This chapter is intended to provide an overview of these three approaches by discussing their background development and key concepts. As noted in the introductory chapter, this particular work is influenced primarily by the Montreal School approach and so, the chapter will discuss this approach in greater detail as compared to the other approaches. The final section provides a brief overview that compares and contrasts the three perspectives and how they approach communication, theorizing, agency, and materiality. Like Brummans et al. (2014), however, I acknowledge the challenge in characterizing CCO approaches: such an endeavor is "fraught with difficulties, since any mapping exercise implies representing reality from a point of view …" (p. 173).

The Montreal School of Organizational Communication

The "Montreal School" approach is based on the work of a group of scholars working out of the Department of Communication at the University of Montreal, Canada. This department was founded in 1987 by James Taylor, now a Professor Emeritus (Brummans et al., 2014). Among this group, François Cooren, a former student of Dr. Taylor's and a professor of communication at the University of Montreal, is a leading scholar, with other department scholars, former students, and others actively publishing broadly from within the Montreal School approach. Generally speaking, the Montreal School approach examines how organizations are communicatively constituted through a process that involves texts, conversations, and conversations about those texts (i.e., metaconversation) and other conversations.

Even though the Montreal School has been treated as a singular group and CCO approach, there are differences across the specific emphases of TMS scholars. Before describing the details of this perspective, key background or precursor theories will be described.

Precursor Theories

There are several theories and scholars that have influenced the Montreal School approach. Bisel (2010) described the quantity of background theories as "dizzying." Taylor and Van Every (2000) broadly identified ethnomethodology and critical theory as precursor approaches for the development of their CCO perspective. In addition, other important theories have helped to lay a basis for the Montreal School approach. Brummans et al. (2014), while recognizing the diverse background of theories as substantial, are more direct in describing the key influences as based on the work of French sociologist Bruno Latour, French semiotician Algirdas Julien Greimas, French philosopher Jacques Derrida, and French philosopher Jean Paul Ricouer. This review highlights some of the key precursor theories or groupings of theories for the purpose of providing background: discourse and interactional approaches, Latour's (2005) Actor-Network Theory, and Algirdas Greimas' (1987) narrative approach. While there are other influences, the aforementioned provide a beginning basis for understanding the distinctive and key assumptions of the Montreal School approach to CCO.

Ethnomethodology was initially developed by sociologist Harold Garfinkel (1967). Garfinkel's ethnomethodological approach was developed in response and critique to Talcott Parson's sociological approach that focused on how people internalized social rules. Garfinkel argued that such an approach treated people as "judgmental dopes" who lacked free will. Instead, Garfinkel emphasized the importance of interactions in everyday life and how, in everyday interactions, people made their behaviors "accountable" to others. Moreover, in these accounts, people "constituted" their actions. For example, a classroom at a university with people gathered in it does not necessarily mean that a class is being taught. However, when an individual walks into the room, utters a greeting, and begins to talk about what will be occurring in class, then the situation as a class can be distinguished from other activities such as a club meeting. In addition, others in the room participate in the communicative constitution of the scene as a classroom: they construct their identities as "students" through their communication by taking notes and indicating when they wish to speak by raising their hands and being called upon by the person speaking from the front of the room. Taylor and Van Every applied Garfinkel's approach to organizations in arguing that accounts constitute organizations as "self-organizing system[s]" (p. 11).

While Garfinkel's ethnomethodological approach highlights the importance of interactions in constituting systems such as organizations or society, the work of *conversation analysts* (CA) such as Harvey Sacks, Emmanuel Schegloff and Gail Jefferson provided a way of documenting and analyzing the minutiae of interactions through the detailed transcribing of conversations (see Sacks, Schegloff, & Jefferson, 1974). What these scholars demonstrated was how conversations are "locally" and "jointly" managed in interactions.

Deidre Boden's (1994) *The Business of Talk* was especially important in applying the insights from ethnomethodology and conversation analysis to organizations. Boden added the idea of "*lamination*." Organizational communication does not occur in a single conversation, but rather in and through multiple, inter-related conversations. These conversations together function to constitute the organization.

The text-conversation perspective draws on French sociologist of science and philosopher Bruno Latour (2005) and what is known as *Actor-Network Theory* (ANT). As a sociologist, a key question that Latour addressed was what is the "social"? Latour argued that explanations of the social that referred to concepts such as structure and seemingly invisible forces were inadequate in that they ultimately did not explain the social but rather described an outcome of the social. ANT examines how social groupings are formed through *associations*. Latour proposed that collectives are formed when one member associates with another, creating a network. It is these associations, rather than external forces that create social groupings. In applying ANT to CCO, scholars have examined how associations are formed through how people talk about connections across individuals and/or across entities labeled as organizations.

The text-conversation perspective is also known as a narrative approach in focusing on communication as story. TMS has been strongly influenced by *Algirdas Greimas' Narrative Theory*. Greimas' theory describes narrative as consisting of the following components: there is a breach or disruption that occurs, the protagonist of the story goes on a quest, along the way, s/he is helped by "helpers" or may be blocked by "blockers" in the quest (see Marsen, 2014 for an application of Greimas' narrative approach to crises). Cooren (2001) utilized the *Star Wars* movie saga to illustrate how to apply Greimas' narrative perspective. While any of the *Star Wars* movies could be analyzed from a narrative perspective, Cooren focuses on the original, *New Hope*. In this story, the robot droid, R2D2, and then eventually, Obi-Wan Kenobi and Luke Skywalker are tasked with getting the plans for the Death Star to the rebels. Along the way, they are *helped* by individuals such as Han Solo, and *blocked* by others such as Darth Vader and Storm troopers.

To illustrate how to apply Greimas' narrative approach to an organization, I focus on a university organization, hypothetically called University A. University A is a primarily undergraduate-serving public institution that has as its primary mission educating students. University A is currently being *blocked* through a state-budget cut, sagging enrollments, and a nearby competitor university with a lower-tuition. However, a grateful alumni with a healthy bank account decides to donate a large sum of money to the university that *helps* to fund scholarships to attract and support students. Greimas' narrative approach highlights and describes the different organizational dynamics that are involved with the development and evolution of organizations.

Together, ethnomethodology, conversation analysis, actor-network theory, and Greimas' narrative approach provided a foundation for the development of the Montreal School approach. However, there are other theoretical perspectives that have influenced the Montreal school approach. Therefore, this overview is partial. The background theories that were reviewed illustrate the interactional and discursive-orientation that TMS has in conceptualizing the communicative constitution of organizations. In addition, these approaches contribute to explaining the view of agency and the role of non-human agency in the TMS perspective.

Within the Montreal School approach, there are several frameworks that have been developed to explain how organizations are communicatively constituted. The following sections highlight James Taylor's work on texts and conversations and François Cooren's (2010) ventriloqual approach.

Texts, Conversations, and Meta-Conversations

Taylor observed that organizations consist of several forms of communication that include written forms (*texts*) and spoken forms (*conversations*). As Taylor (2009) explained of the distinction between conversations and texts:

> *conversation* [is] to-and-fro construction of a collaborative fabrication of *shared talk* [emphasis added], and *text* [consists of] forms of communication that may be extended in time and space beyond the bounds of a single time/space setting. … (p. 157)

Conversation can be thought of as communication as it is happening between two or more people talking in real-time whereas text is communication that endures over time and space as in a written document or policy. A conversation may become a text as people talk about the conversation or if the conversation is recorded in minutes to be remembered later on.

When people talk in organizations and perform various actions, it is in a world filled with conversations, texts, and conversations about those texts and other conversations. For example, during a customer-interaction, an employee may say to a customer (the conversation) that the business has a 30-day satisfaction guaranteed policy (the text).

There is an additional layer here in that when people talk, it may also be *about* prior conversations and about texts. This communication *about* communication is referred to as *metaconversation*. For example, Robichaud (1999) examined a public discussion in which the residents of a community had an opportunity to talk with, and complain to, their local mayor. During the course of the discussion, the mayor referred to past conversations within his political party and among the constituents who have expressed specific complaints in the past. It is through these references to other communication that a particular story of past actions and of the city government as an organization is developed. In this respect, the city government is communicatively constituted.

Another key idea within the Montreal School is that of "*translation*." In everyday usage, translation usually refers to taking something that was stated in one language, and restating it in another language (e.g., from "hello!" to "bonjour!" in an English to French translation). However, the term "translation" is used in a different sense in the Montreal school that is more akin to the idea of "transformation." A translation involves a change "both in medium and in form" (Brummans et al., 2014). In parallel manner to the four flows model that will be explained in the following section, there are four "translations" described through the Montreal School approach.

The first translation involves the "*organization as a network of practices and conversations*." As Brummans et al. (2014) explain, "the term *network of practices* refers to the situated activities (and their associated conversations) that accomplish the work of the organization" (p. 177). What is specific about this network as organizing activity is that there is a common *coorientation* among multiple parties to a single, shared "object." As an example, Brummans et al. describe a hospital which consists of physicians, nurses, administrators, etc. These different groups share a common orientation to treating patients. In order to treat patients, there is a set of *situated activities* that must occur that include scheduling appointments, registering patients, talking with patients, potentially running medical tests, diagnosing ailments, and treating illnesses, among other activities. Each of these activities also involves conversations in some form or other.

The second translation involves "*organization as mapping collective experience through distanciation*." Distanciation involves "the transporting of an organization across space and time" (Putnam & Fairhurst, 2015, p. 378). How does this occur? When people speak of something that occurred somewhere else at some other time, distanciation occurs. For example, when someone begins a meeting by stating, "at our last meeting, we discussed ideas for this year's fund-raiser ...," she is bringing a past conversation into the present. Note that this process occurred through *metacommunication* or communicating about communication in talking about a past meeting conversation.

The third translation is explained as "*organizing as authoring the organization and its purposes through textualization*" (Brummans et al., 2014, p. 170). This translation emphasizes how conversations are turned into texts whether they are referred to at a future point (as the preceding hypothetical example illustrates) or whether they are turned into written documents. For instance, in the preceding example, the speaker could have just as easily referred to meeting minutes: "according to the minutes, we discussed ..." The organizational world is one filled with texts such as minutes, policies, mission statements, vision statements, strategic plans, etc. These texts represent conversations. In turn, these texts are used as a basis for organizing and for future actions of organizational members.

The fourth translation is explained as "*organization as representation and presentification*" (Brummans et al., 2014, p. 180). This translation highlights how the process of organizing involves how the organization achieves an identity that is treated with authority. For instance, the opening of the introductory chapter highlighted a comment made during a press conference on how a particular decision was an "organizational decision." In describing this as an organizational decision, the voices of the many who participated in the decision are summarized in a unified way (whether or not they agreed with the decision or participated equally in the decision). This authority is also brought into and made relevant in the present through referring to the "organizational decision."

Agency

One key implication of TMS is that not just people, but objects and texts can perform actions. The Montreal School approach subscribes to a "post human" or "nonhuman" view of agency. To understand how this can be, it is important to note that TMS draws from Latour's notion of agency as anything that "makes a difference" in the performance of an action. When agency is viewed in terms of performance rather than in terms of intentions or goals, it becomes possible to recognize that nonhumans are agentic or have agency. In other words, something does not have to think in order to make a difference in a situation.

In order to understand this, we must look at what agency means from this perspective. When agency is viewed as someone or something that makes a difference in a given situation, it is being viewed as a relational activity (see Robichaud, 2006). Agency therefore is not a property that an individual has in isolation from others. Rather, the degree to which someone or something exerts agency depends on a given situation and on how others respond. Agency therefore is performed. As a result, *anything*, including objects, can be said to have agency. Humans can make a difference in a given situation through their talk and/or actions. However, nonhuman entities such as a note posted on a staff room door can make a difference. As an example, Cooren described how a road speed bump is agentic in making a difference in action by signaling that drivers should slow down. As Cooren explains:

> artifacts have a big role to play in the communicative constitution of an organization. They matter a lot. They count. They display agency to the extent that they "make a difference." They communicate how an organization is perceived and experienced—think of buildings, machines, and logos, for instance. They express their missions, official positions, and policies—think of texts and web sites. (Schoeneborn et al., 2014, p. 293)

This can be seen in everyday talk when an organizational member claims that something must be done because it is in the rule or policy book.

While the Montreal School differs from the other CCO approaches in how nonhumans are involved in organizational communication, for the Montreal School, humans do still play an important part in organizational communication. As Cooren explains: "the human actor is an obligatory passage point" (from Schoeneborn et al., 2014, p. 299).

Ventriloqual Theory

A specific theory developed within the Montreal school is François Cooren's *Ventriloqual Theory* (Cooren, 2010). Thinking first about ventriloquism as a form of stage entertainment, there is a doll known as a "dummy" or "vent" that is controlled by a ventriloquist who makes it appear as though the dummy is talking. As part of a comedy show, the ventriloquist makes it seem as though the dummy has a mind of its own, and when interacting with the dummy on stage, others talk directly to the dummy. Cooren argues that this type of communication activity occurs in everyday life and in organizations. When an organizational member says, "according to our policy …" or "according to the minutes …", he is acting the part of the "ventriloquist dummy" in speaking on behalf of the organization.

Cooren's Ventriloqual Theory is important in addressing the issue of how are organizations communicatively constituted. In the Ventriloqual perspective, communicative constitution occurs as organizational members speak on behalf of other beings that can include people (e.g., a boss), texts (e.g., organizational policies), or other entities (i.e., the "organization"). In Cooren's (2010) seminal book on Ventriloqual Theory, *Action and Agency in Dialogue: Passion, Incarnation and Ventriloquism*, he describes how action and passion are interconnected by first examining the etymological root of passion as being connect to "passivity" in that passion implies that one is passive to or subject to another. As Cooren elaborates:

> any action (activity) involves in one way or another a form of passion (passivity) to the extent that any interlocutor can position herself or be positioned as animated, moved, motivated, led, inspired, enthused, stirred, forced constrained, affected by specific beings (to be identified). (p. 58)

Ventriloqual Theory highlights two "directions" of action: upstream and downstream. The former direction of being "upstream" highlights "not only what appears to bring into being interlocutors' behavior and action, that is, what *animates* them, but also what these interlocutors make present, that is, represent in their interactions" (p. 57). Whereas the "downstream" direction involves "the signs that interlocutors produce and what they do or make interlocutors do" (p. 57). In other words, that which is upstream refers to what an interlocutor attributes his or her actions to (e.g., "according to our policy …") while that which is downstream refers to what the interlocutor does or says (e.g., "… we are not providing this service today").

Another key relevant concept for Ventriloqual theory is the notion of presentification (Benoit-Barné & Cooren, 2009). As Benoit-Barné and Cooren explain:

> we rely on the term presentification to signify those ways of speaking and acting that are involved in making present things and beings that, although not physically present, can influence the unfolding of a situation. In other words, making something or someone present dislocates any interaction, allowing the transportation in time and space of different agents (collective and individual, human and nonhuman) in a given interaction. (p. 10)

Ventriloquilism operates through presentification or bringing another (person, text, etc.) into the present "here and now" of an interaction. In doing so, the "being" of the organization or other relevant authority figure is made relevant in an interaction.

To illustrate the above ideas, Benoit-Barné and Cooren (2009) describe an interaction that Cooren observed at a Médecins Sans Frontières (also known as "Doctors Without Borders") medical facility when a regional coordinator was touring the facility with Cooren. A note had been left for the technicians regarding appropriate sterilization techniques and processes. Upon entering the room where the note had been posted, a technician remarked, "in any case, maman, we thank you for the note. [showing it on the wall] It has—it has been honored" (p. 22). The note functioned to "presentify" the wishes of the regional coordinator and operated to "ventriloquize" the regional coordinator's authority as well as that of the Médecins Sans Frontières organization.

Summary and Critiques

While TMS is robust in terms of providing several ways to describe and analyze organizational communication and how organizations are communicatively constituted, there are several critiques of this approach. First, it does "not provide a coherent, unified perspective" (Brummans et al., 2014, p. 182). As this review highlights, there are several, inter-related ideas developed by TMS scholars that themselves can be used as separate frameworks and theories. In addition, there are difference among CCO scholars themselves. This is not necessarily a flaw as intellectual diversity is important for the development of knowledge, rather, these differences just present challenges in characterizing the approach. Second, perhaps due to the diverse range of theories written in different languages that have influenced this perspective, the ideas of TMS are complex and difficult to understand. Third, some of the claims to TMS are controversial, especially claims dealing with arguing that non-humans can have agency.

There are a number of strengths of the TMS perspective. First, in focusing on discourse and language in social interaction, TMS provides an important focus on communication activities and in understanding the processes of how those activities function to constitute organizations. Second, it is international in its origins and in its scholarly community. Such breadth of reach has afforded a robust scholarly conversation where the ideas of TMS have been applied and extended in many different ways.

Four Flows Model

The Four Flows model was developed by McPhee and Zaug (2000). McPhee and Zaug's key article was published in 2000 in the *Electronic Journal of Communication* and formed the basis for the National Communication Association 2002 Pre-Conference on the CCO approach. This article also formed the starting point for Putnam and Nicotera's (2009) edited book collection, *Building Theories of Organization: The Constitutive Role of Communication.* McPhee and Zaug use "flow" to refer to forms of messages as process. According to these authors, organizations are constituted through four interrelated flows. The key precursor theory for the development of the Four Flows model is Structuration theory.

Precursor Theory: Structuration

The Four Flows model is based on the work of British Sociologist, Anthony Giddens and his work on developing Structuration Theory (see Giddens, 1984). Giddens describes how within society, there is a tension or "duality" between structure and agency. Structure may be conceptualized as external forces that guide human behavior. Agency refers to individual, human choices. A challenge in social theories that Giddens and others have noted is that theoretical explanations that focus on structure are unable to account for individual choices and differences. As important leaders such as Mahatma Gandhi and Martin Luther King, Jr. have shown, individuals *can* make a difference in challenging and changing the status quo.[1] Conversely, social theories that focus on the role of the agent have difficulty in accounting for the constraints that individuals seem to operate under. In other words, people are not able to make any choice that they want to (i.e., not "anything goes"). According to Giddens, both structure *and* agency exist and work together in a relationship of "duality."

Society is not just based on structures dictating what people do as people have free will. Alternatively, people do not just get to pick and do whatever they want to do. Rather, there are constraints and expectations placed on their actions. Giddens instead advocated for a "duality of structure" with *both* structure *and* agency operating together. People can and do make choices, but they do so within constraints. In the Four Flow model, McPhee and Zaug identified four "flows" that function to constitute or construct organizations.

The Four Flows

In the Four Flows model, an organization is defined as:

> a social interaction system, influenced by prevailing economic and legal institutional practices, and including coordinated action and interaction within and across a socially constructed system boundary, manifestly directed toward a privileged set of outcomes. (McPhee & Zaug, 2009, p. 28)

A flow refers to a field of messages, circulating system, or message processes. Thus, a flow consists of communication messages related to a particular topic or issue. McPhee and Zaug argue that within organizations, there are four important flows that function to constitute organizations: membership negotiation, organizational self-structuring, activity coordination, and institutional positioning.

Membership Negotiation. Organizations consist of individual members. However, how membership occurs is part of a complex communication process. First, individuals must join an organization, usually through a hiring process if the individual is joining the organization as an employee. However, being hired is not sufficient for new members to then know how to act and behave as an organizational member. There is typically a socialization process where individuals learn of both the formal and informal rules as well as the culture of the organization. What these concepts highlight is that membership in an organization is not a one-time event, but an ongoing informal negotiation process that occurs whenever organizational members are interacting. Negotiation here is not meant in the formal sense of explicitly discussing an agreement, contract, or policy. Rather, the term "negotiation" points to the general idea that there is interaction involved with figuring out and "doing" the role of employee.

Organizational Self-Structuring. Organizations consist of a grouping of people, however, not all groups are organizations. This is the case if we consider what is different about a crowd at a street fair as compared to the cafe and its employees down the street. Organizations are distinct from other groupings of people such as crowds or mobs in that they have a structure that functions to designate expectations, responsibilities, and decision-making.

However, structure does not just happen. Rather, organizations are *self-structuring* in that the structure is the result of communication processes. Also, organizational structure is not just about work processes but includes "the internal relations, norms, and social entities that are the skeleton for connection, flexing, and shaping of work processes" (McPhee & Zaug, 2009, p. 36). McPhee and Zaug provide the following examples of communication that provide for organizational self-structuring:

> charters, organization charts, policy and procedure manuals; decision-making and planning forums; orders directives, and the more casual announcements that often substitute for them; processes of employee evaluation and feedback; budgeting, accounting, and other formalized control processes. (p. 36)

Activity Coordination. Organizations typically do not exist just to exist; they have purposes and goals towards which the activities of its members are directed toward achieving. Activity coordination refers to these actions and processes of organizational members in accomplishing the work of the organization. As McPhee and Zaug (2009) explain, "the process of adjusting the work process and solving immediate practical problems requires the sort of communication we call activity coordination" (p. 38).

For an example, McPhee and Zaug described a study of a Paris metro control room and the activity involved during a train breakdown. This activity involved gathering information about the breakdown, centralizing information flow and decision-making about the breakdown with the initial controller who received the information about the breakdown, and anticipating delays. Activity coordination highlights how people do things in organizations, and organizations are constituted when these actions or activities are coordinated together in fulfilling the purpose and goals of the organization.

Institutional Positioning. Organizations are not "islands" that exist on their own in society: they are interconnected with society and with other organizations. At a basic level, organizations may depend on suppliers, buyers, funders, government regulations, etc. This interdependence is both practical *and* relational; McPhee and Zaug (2009) explain that "organizations must constitute themselves as practical relational partners" (p. 41).

Organizations are institutional in the sense that there are "organization-sets" of peer and related organizations that have certain things in common such as structures and regulations. For example, there is a general structure and expectations for organizations such as retail organizations as compared to hospitals. Retail organizations are expected to sell goods, whereas hospitals are expected to treat patients. Institutional positioning enables organizations to exist without having to wholly invent or re-invent themselves.

Summary and Critiques

The Four Flows Model is significant in paving the way for the development of theories of CCO. However, as with all theories and models, there are critiques, many of which have been issued from scholars of the Montreal School.

First, in examining the Four Flows model, Taylor (2009) asks, "what exactly is a 'flow' …?" (p. 154). According to Taylor, how flow is defined and what it means is vague. For instance, a flow could consist of communication interactions, episodes, patterns of activities, a history (Taylor, 2009); which, if any, of these describes "flow" is unclear in the model proposed by McPhee and Zaug.

Second, another shortcoming identified by Taylor (2009) is that "their program of research needs to be supplemented with a more precise theory of communication that has as its objective to trace the genesis and grounding of organizational form and process in the communication event" (pp. 154–155). In other words, it does not sufficiently focus on communication as a practice. This particular critique, however, is also viewed as a strength by other scholars.

Putnam, Nicotera, and McPhee (2009) described four implications of the Four Flows Model. First, it views communication as "wholly constitutive of organizations" (p. 11). Second, it emphasizes "contextualist" over "structuralist" thinking, meaning, it focuses on organizational processes over an abstract structure. Third, structure and agency are interconnected in a duality. And finally, it "resists the temptation to take a micro-level position" (p. 11) by examining communication across different "episodes" or settings.

The Four Flows model emphasizes a *deductive* rather than *inductive* approach to organizational communication (Brummans et al., 2014). The reasoning for a deductive approach is because of the view that to understand organizations, one must look beyond single conversations or episodes. Rather, organizations extend beyond these in their complexity. McPhee describes the Four Flows model as a meta-theory: "The Four-Flows distinctions seem, as we hope, to valuably cut across meta-theories—they could be, for instance, four dimensions of imbrications or actant-relations exhibiting ventriloquism in varied ways" (Schoeneborn et al., 2014, p. 294). In some views the Four Flows theory is viewed as a meta-perspective that in turn can encompass or subsume other CCO approaches.

Luhmann's Theory of Social Systems

Niklas Luhmann was a German sociologist who lived from 1927 to 1998. The core of Luhmann's social theory was communication. He wrote prolifically about social systems, hence, it is logical that Luhmann's work be applied to organizational communication. For Luhmann, organizations are one of three types of human systems (the others consisting of interactions and society more broadly).

Luhmann developed a specific systems theory approach also referred to as the theory of social systems or TSS (Schoeneborn, 2011). There are many different types of systems theories that have been developed. One specific characteristic of Luhmann's systems approach is that systems "are marked by their capacity to produce *by themselves* a difference that distinguishes them from their own environment" (Brummans et al., 2014, p. 184). Luhmann referred to this capacity as being *self-referential*. A system is something that can distinguish itself from its environment. Another way to explain this is that a system is something that can create a unique and distinctive identity. As Schoenenberg (2011) explains, "social systems such as organizations fundamentally emerge by the distinction between the system and its environment" (p. 678).

A key component of Luhmann's theorizing is communication. Luhmann's emphasis on communication systems is so strong that he emphasizes that people are a part of communication systems rather than the converse that communication is a part of human systems: "Accordingly, social systems are not comprised of persons and actions but of communications" (Luhmann, 1989, p. 145, cited in Schoeneborn, 2011, p. 670). For Luhmann (1995), communication consists of three components: information, utterance, and understanding. Information consists of the *what* of communication or content. Utterance describes the *how* and *why* of communication. Understanding both distinguishes and bridges information and utterance.

Another key aspect of Luhmann's theory is *autopoesis* or the idea that a system regenerates itself. This is a concept that Luhmann uses from the work of biologists Maturana and Varela (1987). For Luhmann, social systems regenerate themselves through their communication. As Schoeneborn (2011) explains, "the organization only exists as long as it manages to produce further communications, which call forth yet more communications" (p. 670).

Luhman identifies three different social systems: interaction (face-to-face), organizational, and societal. These can also be described as the micro-level (interactional), the meso-level (organizational), and the macro-level (societal) level systems (Brummans et al., 2014). With the organizational level, along with communication, decision is a key aspect of organizations. Decisions or "decision communication" is what distinguishes organizational communication from mere interaction (Brummans et al., 2014). For Luhmann, "decisions are communications" (Luhmann, 2003, p. 32, cited in Schoeneborn, 2011, p. 671). Schoeneborn (2011) further explains, "the organization comes into being whenever speech acts adopt the form of decisions" (p. 671).

Summary and Critiques

Luhmann's perspective has a great deal of relevance for organizational communication scholars given that he addresses organizational systems and emphasizes communication. Luhmann's approach foregrounds the role of communication in constituting (and re-constituting) systems through the concept of *autopoesis*. His ideas regarding decision communication are provocative in calling attention to specific communication that makes organizational communication *organizational*.

However, Luhmann's approach has not received as much attention as the other CCO approaches described in this chapter. There are several reasons for this. Luhmann's ideas are difficult to understand and are written in ways that do not make the ideas any easier to understand. This is a critique that was recognized by Luhman in that he intentionally did not want his theory to be easily understood as he wanted to cause a paradigm shift in how social systems are understood. Moreover, his writing was initially done in German with some of his work being translated into English, thus limiting the reach of his work to a broader audience. Although, some work has been done to help translate Luhmann's ideas to share with English-reading audiences (see Schoeneborn, 2011). In addition, his work was introduced more recently to U.S. organizational communication scholars compared to the longer history of the Four Flows and Montreal School perspectives. Another critique of Luhmann's perspective is that he overly focuses on decisions in organizations. Finally, Schoeneborn (2011) also identifies a limitation of Luhmann's perspective in "neglect[ing] the role of material agency in the communicative construction of organizations" (p. 663).

Luhmann's notion of contingency is especially relevant for the overall project of this book in applying a CCO lens to crises. As Schoeneborn (2011) explains, "'contingency' means an instance of 'it could be otherwise' and thus represents potentiality as opposed to actuality" (p. 672). Some of the crises examined in this work are instances that "could have been otherwise." Overall, Luhmann's perspective holds a great deal of promise in being applied to contemporary problems and in understanding how organizations are communicatively constituted.

Comparing and Contrasting Across Perspectives

This section compares and contrasts the three main CCO perspectives across three key areas: views of communication, agency, and materiality.

Communication

In the CCO perspective, communication is not merely about the transmission of information or expression of ideas. It is a social, meaning-making process in which symbols and objects participate in this process. The three CCO perspectives define communication in slightly different ways. For the Montreal school, communication is *action* and *transactional* (see Schoeneborn et al., 2014). Also, communication goes beyond humans but also nonhumans are participants in communication activities.

The Four Flows model holds similar views of communication as the other views, but emphasizes the symbolic nature of communication. In the Four Flows model, communication is interaction that occurs within a context.

In Social Systems theory, communication is a "form of observation" that consists of information, utterance, and understanding (Schoeneborn et al., 2014). Information refers to the content of *what* is said. Utterance refers to the activity of communicating or *how* something is said as well as *why* it is said. Understanding involves being able to distinguish what is said (information) from how and why it is said (utterance). Thus, one key difference between Luhmann's approach and the other two CCO approaches is that communication is not conceptualized as "action" given that the idea of "understanding" is not captured through the notion of "action" (Schoeneborn et al., 2014).

Agency

Brummans et al. (2014) explain the contrasting views on agency between the three main CCO approaches as follows: the Four Flows approach focuses on human agency; the Montreal School approach includes humans and nonhumans as agents; and the Social Systems approach focuses on communication as the "author" of organizational systems.

Agency is an important area of study in organizational communication, and an important area where the CCO approaches diverge. In basic terms, agency can be thought of as the ability to do something. The tension in conceptualizing agency though has to do with where agency is located. For the Four Flows model, agency is located in individuals and the ability of individuals to make intentional choices. As such, agency is a distinctly human capacity because only humans can be intentional in their choices.

Alternatively, for the Montreal School, agency is defined both strictly yet loosely as the capacity to make a difference. Because of this close connection between agency as a form of action, humans *and* nonhumans or inanimate objects such as written texts, post-it notes, photocopiers, etc., can be said to be agents and have agency in that these objects can "make a difference" in actions.

Because of the Social Systems approach's emphasis on communication processes, this perspective can also be said to de-emphasize human agency. However, it is also an approach that has not paid as much attention to materiality. Therefore, while it shares some views on agency in common with the Montreal School, it is not fully aligned with its views on agency.

Materiality

Among the three main CCO approaches, the Montreal School has the greatest emphasis on materiality, especially given that in TMS, nonhumans may exert agency. In contrast, TSS focuses mainly on interaction rather than materiality in its description of organizations (Schoeneborn, 2011).

The Montreal School's approach to materiality can be explained through Cooren's (2010) development of Ventriloqual Theory. Cooren's (2015) approach to materiality is based on understanding what the term material means based on its linguistic/word origin. When something is material, that means that it matters in some way in a particular situation and/or to specific people. When something matters, it supports or "stand behind" something else. Thus, materiality is not just an inherent property of "things" based on their physicality; rather, materiality is based on how entities, substances, or phenomena relate to each other.

Summary

While the three CCO approaches differ (Bisel, 2009), they have also been in collegial dialogue with each other as part of the overall project in advancing a constitutive perspective of organizations (e.g., Schoeneborn, 2014). In the application chapters to follow, I dominantly rely on concepts and assumptions from the Montreal School. This emphasis is driven in part because of how the Montreal School's focus on language-use in social interactions can provide valuable insight on one's own communication practices in "conversations." As the case studies will illustrate, it is often in organizational conversations, meetings, emails, and interactions that a great deal of important CCO work can occur. The following chapters will also illustrate how conversations can have an important impact on how a crisis unfolds and how it is addressed.

As students of organizational communication and as participants in various forms of organizational communication, we can be empowered to participate in organizational conversations where we can make deliberate, conscious choices regarding our communication practices.

Table 2.1. Summary of Three Communication as Constitutive of Organizing Perspectives.

Perspective	The Montreal School	The Four-Flows Model	Social Systems Approach
Summary	Communication is the 'site and surface' of organizing. Organizations are constituted through the association of texts, conversations, and metaconversations.	Organizations are constituted through four interrelated flows: membership negotiation, organizational self-structuring, activity coordination, and institutional positioning.	Organizations are one (of three) types of social system. Systems are: self-referential (able to distinguish themselves from their environment), autopoetic (regenerative), and consist of communication.
Key Current Theorists	François Cooren	Robert McPhee	Dennis Schoeneborn
Early Theories and Theorists	James Taylor; language and social interaction approaches (e.g., ethnomethodology, conversation analysis); critical theory; Bruno Latour's Actor-Network Theory; Algirdas Greimas' Narrative perspective; Jacques Derrida; Jean Paul Ricouer	Anthony Giddens' Structuration Theory	Niklas Luhmann's Social Systems Theory
Approach to Communication	Communication is action and transactional. Focuses on associations that occur	Communication is symbolic; it is interaction within a context.	Communication consists of information (what is said), utterance (how and why something is said),

(*Continued*)

Table 2.1. (*Continued*)

Perspective	The Montreal School	The Four-Flows Model	Social Systems Approach
	through interactions. Communication as stories (narratives).		and understanding (distinguishing between information and utterance). Decision communication is what makes organizational communication distinct.
Theorizing	Several theories and concepts within TMS such as text-conversation, ventriloquism, and four translations	Deductive approach that can be used as meta-theory to encompass other CCO approaches	Paradigm shift in how social systems are conceptualized
Agency	The capacity to make a difference. A property of an entity based on its relevance for a given situation, therefore, humans and nonhumans (e.g., texts, objects, machines) can be said to have agency.	Only humans have agency. Structure and agency connected in a 'duality.'	Communication is the 'author' of organizational systems. Agency is not necessarily human.
Materiality	Something that matters to a given situation.	Not a direct focus of this perspective.	Focuses mainly on interaction and neglects materiality.
Criticisms	Too complex. Distorts and misapplies the concept of agency.	Flow is vaguely defined. Limited view of communication as practice.	Difficult to understand.

Note

1. This is not meant to imply that individuals alone make a difference given that the aforementioned individuals were a part of a social movement working with others to make a difference.

References

Benoit-Barné, C., & Cooren, F. (2009). The accomplishment of authority through presentification: How authority is distributed among and negotiated by organizational members. *Management Communication Quarterly*, *23*(1), 5–31. https://doi.org/10.1177/0893318909335414

Bisel, R. S. (2009). On a growing dualism in organizational discourse research. *Management Communication Quarterly*, *22*, 614–638.

Bisel, R. S. (2010). A communicative ontology of organization?: A description, history, and critique of CCO theories for organization science. *Forum Management Communication Quarterly*, *24*, 124–131. https://doi.org/10.1177/0893318909351582

Boden, D. (1994). *The business of talk: Organizations in action*. Cambridge: Polity.

Brummans, B. H. J. M., Cooren, F., Robichaud, D., & Taylor, J. R. (2014). Approaches to the communicative constitution of organizations. In L. L. Putnam, & D. K. Mumby (Eds.). *The SAGE handbook of organizational communication: Advances in theory, research, and methods* (3rd ed., pp. 173–194). Thousand Oaks, CA: Sage.

Cooren, F. (2001). Translation and articulation in the organization of coalitions: The Great Whale River case. *Communication Theory*, *11*(2), 178–200. https://doi.org/10.1111/j.1468-2885.2001.tb00238.x

Cooren, F. (2010). *Action and Agency in Dialogue: Passion, incarnation and ventriloquism*. Amsterdam: John Benjamins.

Cooren, F. (2015). In medias res: Communication, existence, and materiality. *Communication research and practice*, *1*(4), 307–321.

Garfinkel, H. (1967). *Studies in ethnomethodology*. Englewood Cliffs, NJ: Prentice-Hall.

Giddens, A. (1984). *The constitution of society: Outline of the theory of structuration*. Cambridge: Polity.

Greimas, A. (1987). *On meaning: Selected writings in semiotic theory* (P. J. Perron & F. J. Collins, Trans.). Minneapolis, MN: University of Minnesota.

Latour, B. (2005). *Reassembling the social: An introduction to actor-network-theory*. New York, NY: Oxford University.

Luhmann, N. (1995). *Social systems* (J. Bednarz, Jr. & D. Baecker, Trans.). Stanford, CA: Stanford University.

Marsen, S. (2014). "Lock the doors": Toward a narrative-semiotic approach to organizational crisis. *Journal of Business & Technical Communication*, *28*(3), 301–326. https://doi.org/10.1177/1050651914524781

Maturana, H. R., & Varela, F. J. (1987). *The tree of knowledge: The biological roots of human understanding*. Boston, MA and London: Shambhala.

McPhee, R. D. (1988). Vertical communication chains toward an integrated approach. Management Communication Quarterly, 1, 455–493.

McPhee, R. D., & Zaug, P. (2000). The communicative constitution of organizations: A framework for explanation. *Electronic Journal of Communication*, *10*(1/2), www.cios.org/getfile/McPhee_V10n1200

McPhee, R. D., & Zaug, P. (2009). The communicative constitution of organizations: A framework for explanation. In L. L. Putnam & A. M. Nicotera (Eds.), *Building theories of organization: The constitutive role of communication* (pp. 21–47). Madison, NY: Routledge.

Putnam, L. L., & Fairhurst, G. T. (2015). Revisiting "organizations as discursive constructions": 10 Years Later. *Communication Theory*, *25*(4), 375–392. https://doi.org/10.1111/comt.12074

Putnam, L. L., & Nicotera, A. M. (Eds.). (2009). *Building theories of organization: The constitutive role of communication*. Madison, NY: Routledge.

Putnam, L. L., Nicotera, A. M., & McPhee, R. D. (2009). Introduction. In L. L. Putnam & A. M. Nicotera (Eds.), *Building theories of organization: The constitutive role of communication* (pp. 1–19). Madison, NY: Routledge.

Robichaud, D. (1999). Textualization and organizing: Illustrations from a public discussion process. *The Communication Review*, *3*(1–2), 103–124.

Robichaud, D. (2006). Steps toward a relational view of agency. In F. Cooren, J. R. Taylor, & E. J. Van Every (Eds.), *Communication as organizing: Practical approaches to research into the dynamic of text and conversation* (pp. 101–114). Mahwah, NJ: Lawrence Erlbaum Associates.

Sacks, H., Schegloff, E. A., & Jefferson, G. (1974). A simplest systematics for the organization of turn-taking for conversation. *Language*, *50*, 696–735.

Schoeneborn, D. (2011). Organization as communication: A Luhmannian perspective. *Management Communication Quarterly*, *25*(4), 663–689. https://doi.org/10.1177/0893318911405622

Schoeneborn, D., Blaschke, S., Cooren, F., McPhee, R. D., Seidl, D., & Taylor, J. R. (2014). The three schools of CCO thinking: Interactive dialogue and systematic comparison. *Management Communication Quarterly*, *28*(2), 285–316. http://doi.org/10.1177/0893318914527000

Taylor, J. R. (2009). Organizing from the bottom up?: Reflections on the constitution of organization in communication. In L. L. Putnam & A. M. Nicotera (Eds.), *Building theories of organization: The constitutive role of communication* (pp. 153–186). New York, NY: Routledge.

Taylor, J. R., & Van Every, E. J. (2000). *The emergent organization: Communication as its site and surface*. Mahwah, NJ: Lawrence Erlbaum.

PART 2

CONCEPTS AND CASES

· 3 ·

PROBLEM IDENTIFICATION AND CRISIS PACING

A crisis is a form of problem. More formally, crisis has been defined as "a major occurrence with a potentially negative outcome affecting the organization, company, or industry, as well as its publics, products, services, or good name" (Fearn-Banks, 2011, p. 2). A key starting point in understanding crisis is to start with understanding the nature of the problem or "major occurrence." The thesis of this chapter is that problems are communicatively constituted. In applying a CCO perspective, this chapter examines how crises themselves are communicatively constituted. A key focus is on the issue of pacing and how the rate at which a crisis unfolds may impact how that crisis is defined and by extension, how people and organizations respond to that crisis. This chapter also addresses the concept of materiality. Even though problems are socially constructed, this should not be taken to imply that they are exclusively, socially constituted. Material circumstances also participate in the construction of crisis.

Because of this chapter's focus on crisis communication, it is especially relevant for those who are interested in applying a CCO perspective to the field of public relations. As such, this chapter begins with a discussion of key ideas in the public relations sub-field of crisis communication, followed by an overview of organizational theories and concepts relevant for crisis construction. The chapter ends with a discussion of a case study that focuses on heatwaves in Europe.

Crisis Communication

There are two ways that the term crisis communication can be explained. First, crisis communication is considered a sub-field of public relations and addresses how organizational members communicate about crises to their external stakeholders such as customers or the general public. This area of crisis communication includes how organizations may respond to a crisis through communication such as explaining, apologizing, and/or remediating (fixing) the crisis. For example, if a company has a faulty product such as a car airbag that does not work properly, it may explain why the problem occurred, it could issue an apology to the public, and it likely will recall its product and pay for the re-installation of properly working airbags. As another example, in 2017, what will likely be the most memorable Oscar Awards mistake occurred when the presenters announced the incorrect movie for receiving the Best Picture award of the year (*La La Land* was announced rather than the correct winner, *Moonlight*). This mistake created a crisis for the accounting firm responsible for handling the award envelops, PriceWaterhouseCoopers. The company's almost immediate response was to take full responsibility for the error, even as it was still investigating the exact cause for the error. The response of accepting responsibility is another way that organizations have responded to crises and illustrates one way that communication and crisis response are interrelated.

Crisis communication may also include how governmental agencies respond during a disaster in the interest of public health and safety. For example, if a tornado is approaching, public warnings are broadcast through mass and social media and tornado sirens turned on. While the areas of organizational communication and public relations overlap in that they both address organizations and communication, they have evolved as separate fields of study and practice.

A second way of explaining crisis communication is more generally as organizational communication *during* a crisis. This area differs from the former area in that it has traditionally examined how organizations communicate *internally* among its members in the process of handling a crisis whereas the former area, as a branch of public relations, addresses how the organization communicates with publics outside of the organization. For example, Bergeron and Cooren (2012) studied how a crisis management team communicates during crisis scenarios. Castor and Bartesaghi (2016) studied teleconferences among Louisiana parish and state emergency managers and federal agency representatives during Hurricane Katrina. In both of the aforementioned projects, the focus was on how team members or inter-organizational representatives communicated with each other during a crisis situation, rather than on how they communicated with the general public. Studies that have examined crisis communication in this way have typically addressed circumstances when the organization handled the crisis poorly.

By definition, the effects of a crisis are severe in their impact on the organization and other stakeholders. One classic example of an organizational crisis is what is known as the Johnson and Johnson Tylenol crisis that occurred in the early 1980s when someone laced Tylenol capsules with cyanide in the Chicago area resulting in the deaths of seven individuals. Johnson and Johnson had to immediately respond through public statements about the tragedies, make decisions on and develop a response strategy, and implement a plan for handling the contaminated pill problem.

According to Fearn-Banks (2011), there are five stages of a crisis: detection, prevention/preparation, containment, recovery, and learning. This chapter, in applying a CCO perspective to crisis management will focus on the first stage of detection and how detection is a communicatively constituted process. One challenge in dealing with a crisis is determining when a crisis is occurring. In applying a CCO lens to crisis, this chapter argues that crises are communicatively constituted. This is not meant to imply that people just make up crises or that the only reason crises happen is because people intend for them to happen. Rather, it is meant to point out that what constitutes a crisis and how material and physical circumstances are made into crises are social-material matters.

Although crises can be severe in their impact, a crisis is a form of problem. The next section provides background information on problems as social constructions as a bridge for understanding how crises are communicatively constituted.

Communicative Constitution of Problems

Problem construction is interconnected with the communicative constitution of organizations vis-à-vis Weick's (1995) work on organizational sensemaking. Weick has been instrumental in organizational communication theorizing by shifting our terminology regarding organizations from nouns to verbs or from "organization" to "organizing." A key aspect of Weick's work is *organizational sensemaking*, a concept that he has utilized to argue for the notion of *organizing* rather than organization given that the former term focuses on process. Weick's work has shown how organizing occurs in situations of uncertainty and ambiguity that are given meaning through communication. This process is summarized through Weick's famous dictum, "how can I know what I think until I see what I say?"[1]

A key point of the argument here is that problems are socially-constructed, a claim that Grint (2005) made in his aptly titled article, "Problems, problems, problems: The social construction of 'leadership'." Grint pointed out that whether or not as well as how a situation is defined as a problem is based on human subjectivity.

Weick's (1995) sensemaking approach addresses how figuring out a problem is challenging to do in the midst of the problem or crisis and that much "sense" of the problem is made after the fact, in retrospection. Weick's sensemaking approach has been applied to conducting post hoc analyses of disasters to examine communication, situational, and perceptual factors that led to a disaster (see Weick, 2001).

Based on the leadership literature, there are three general types of problems: *tame*, *wicked*, and *complex*. The first two types of problems, tame and wicked, are based on the work of Rittel and Webber (1973). A tame problem is one that is complicated but solvable because it is likely a problem that has occurred before and therefore, there is knowledge of what it is and how to solve it based on past experiences. As Rittel and Webber also explain, "there is only a limited degree of uncertainty" (p. 1473) associated with the problem. An example of a tame problem would be planning a heart surgery, developing a timetable for air flights, or training an army. While there is a great deal of risk involved with each of these, set processes and procedures have been developed, tested, and repeatedly practiced to limit the uncertainty of how to carry out these activities. This is not to imply that there is not any uncertainty associated with these, but that the processes of how to navigate these situations are known.

A wicked problem is "complex, rather just complicated" (Rittel & Webber, p. 1473). There is not a clear let alone easy solution. As such, there is not a definitive "stopping point" in that a given solution is likely to generate additional problems. Rittel and Webber also explain that "there is no 'right' or 'wrong' answer, but there are better or worse alternatives" (p. 1473). With a wicked problem, there is a great deal of uncertainty associated with figuring out the problem and possible solutions. Examples of wicked problems have typically included international issues such as the Israeli-Palestinian conflict or the Cuban missile crisis of 1962—these are problems that have been or are still difficult to solve. However, wicked problems can be internally focused as in trying to develop a national healthcare system.

A critical problem is clear in its nature and calls for immediate action (Grint, 2005). In part because of the clarity of the problem, there is little uncertainty associated with identifying the problem and implementing the appropriate solution. Examples of this problem type would include a terrorist attack, a major freeway crash, or job loss. With each of these examples, while the problem is severe, people generally know what to do to respond right away.

Grint makes the point that for a given problem, its category type could change. For example, a tame problem could change into a wicked problem. The example that Grint provides is how a company's financial losses may not become a crisis until the company stockholders decide to sell their holdings en masse. Then, the stockholder actions create a crisis for the company that is more difficult to solve.

In a study relevant for this project, Fairhurst and Cooren (2009) examined the construction of leadership as the "hybrid production of presence" in which "effective" or positively evaluated leadership involved the successful (communicative) mobilization of various agents. Fairhurst and Cooren focused on leaders during crisis situations that were related to weather and natural disaster by comparing and contrasting Louisiana Governor Kathleen Blanco's communication during the 2005 Hurricane Katrina natural disaster with that of California Governor Arnold Schwarzeneggar's communication during the California wildfires.

Another example is taken from Murphy's (2004) study of the January 1998 ice storm disaster in Canada. During this weather event, Quebec experienced a severe ice storm of persistent, freezing rain that "loaded ice onto all surfaces, crushed the electrical system of western Quebec" and caused "essential infrastructure to cease functioning" (p. 256). The damage to the electrical infrastructure of Quebec resulted in the loss of heat and light for two weeks. Food spoiled as freezers lost their power; food sources were also affected when farmers who relied on electricity for maintaining their barns no longer had access to electricity (consider the equipment that is involved with milking cows and keeping barns clean). This was obviously a crisis for Quebec, but the ice storm did not cause a crisis for all in the area.

Murphy contrasted the experiences of the Quebecois with that of the Amish community in upstate New York who were also within the area affected by the storm. The technology of the Amish was akin to the agricultural technology used in 17th-century Europe with no reliance on electricity. The Amish were able to continue heating their homes, cooking their food, milking their cows, and maintaining their barns. Murphy summarized the differences in Quebec versus Amish experiences in the ice storm as follows: "the comparison of Quebec society with the Amish demonstrates that disaster resulted not from freezing rain *per se*, but rather from the vulnerability of the infrastructure that modern society had constructed and upon which it had become dependent" (p. 257). In other words, "susceptibility to disaster was socially constructed, not natural" (p. 257).

By extension, problem construction in organizations can provide several insights into the communicative constitution of organizations in that issues such as what constitutes a problem is interrelated with a constellation of other issues such as situated ideals (i.e., what norms or rules have been violated), agency (i.e., who or what caused the problem), accountability (i.e., who should address the problem), and action (i.e., what agents and resources should be mobilized to address the problem).

The key point is that what constitutes a problem is not self-evident. What may be a problem or crisis for some may be an inconvenience to others. Part of this may be due to how well-equipped a group is to handle a particular situation, and part of this may be due to how a given situation is defined. This is also a matter of how one interprets the material features of a given situation and projects ahead to anticipate what may occur. In addition, the nature of the problem can shift as what initially seems like an easy problem to address may get complex. One challenge in defining a problem relates to the issue of pace or how quickly or slowly a crisis occurs.

Sensemaking and Implications for Problem Construction

Maitlis and Christianson (2014) conducted a comprehensive overview of organizational sensemaking, noting the variety of perspectives and definitions that reflect the diversity of specific theoretical approaches to organizational sensemaking. Based on their review, they presented the following definition for organizational sensemaking as:

> A process, prompted by violated expectations, that involves attending to and bracketing cues in the environment, creating intersubjective meaning through cycles of interpretation and action, and thereby enacting a more ordered environment from which further cues can be drawn. (p. 67)

This definition highlights how sensemaking has typically been conceptualized as a response to the unexpected which then triggers a meaning making process to restore order. This orientation to sensemaking highlights how and why organizational crises have been one significant stream of research in the organizational sensemaking literature (Maitlis & Sonenshein, 2010). Within the area of organizational crises, sensemaking has been applied to understanding what happened and how various factors, including communication, may have played a contributing role to creating the crisis as well as in recovering from crises (see Gephart, 2007; Seeger, Sellnow, & Ulmer, 1998).

Sensemaking itself has been conceptualized as a social construction process where the subjects or topics of sensemaking are viewed as having ontologies that are interconnected with sensemaking processes rather than existing separately from sensemaking. This orientation is emphasized in Weick's (1988) discussion of enactment. Weick breaks down the sensemaking process into three phases: *selection*, *enactment*, and *retention*. In the selection phase, social actors identify cues to which they attend. As implied by Maitlis and Christianson's (2014) definition, "cues" are a key starting trigger for sensemaking processes, particularly when the cues are unexpected or difficult to explain: those cues provide the material for making sense of in the enactment phase.

During enactment, social actors construct the environment, defined as that which seems to exist outside of the organization. Weick, Sutcliffe, and Obstfeld (2005) argued that the environment is not something that is a separate entity from organizational actors during the sensemaking process. The environment is socially constructed during the sensemaking process. One implication of this claim that has also been supported through sensemaking research is that problems or crises themselves are socially constructed (Buttny, 2004; Castor & Cooren, 2006; Grint, 2005; Weick, 1995).

With respect to sensemaking, Castor and Cooren's (2006) project is of the greatest relevance for this project. They examined how a faculty senate made sense of a university's budget crisis by analyzing the accounts that the senators' presented for the crises. They described this process as problem formulation in that the accounts functioned to construct the problem and critique the proposed solution for the budget crisis. Castor and Cooren related this as a process of sensemaking in that the senators were trying to determine what happened during the crisis (defining the problem). This process relates to what Weick described as the process of interpretation. Castor and Cooren's project was intended to contribute to the understanding of *how* interpretation occurs during sensemaking by examining a specific discursive process that occurred during problem formulation. Castor and Cooren described this process as one of selecting an agent in a chain of agencies. In other words, the faculty senators during their discussion constructed an interconnected net of agents, or more appropriately actions and activities (see Czarniawska, 2009). The construction of the crisis was dependent on where in the net the faculty senators located cause versus effect.

Time is an implicit aspect of constitutive processes, in that process implies activity over time. However, time and timing have been addressed more so in describing the past as in bringing the "there and then" into the "here and now" or in describing past processes.

Crises and Pacing

A crisis may occur over a short or long duration. Crises that happen quickly cause a sudden jolt or change that is self-evident. When crises occur quickly, they can also be dealt with in that they may be relatively easy to identify. However, other crises are slower to unfold, and this may also make it harder for them to be identified as crises or potential crises.

McConnell (2003) identified three forms of crisis with respect to their pacing. The first type is a *sudden crisis*, which, as indicated by the name, is quick to occur. Examples of this include mass shootings or an earthquake. The second type is a *creeping crisis*. As McConnell described, "vulnerable conditions and pressures build up slowly, often over many years" (p. 188). Examples of creeping crises, according to McConnell, include global warming and the spread of Bovine Spongiform Encephalopathy (BSE) (also known as mad cow disease) in the United Kingdom. McConnell also includes the September 11 terrorist attack in the U.S. as an example. While the actual attack unfolded over a span of hours, the circumstances leading up to the attack were many years in the making. The third type are *chronic crises*. They may be similar to creeping crises in their slow development, however, they differ in that they are ongoing and also do not have an obvious solution. Humanitarian problems would be an example of a chronic crisis.

The case analysis of this chapter is with creeping crises as they may be difficult to identify because of their gradually developing nature. As an example, consider global warming and the challenges in identifying trends in global warming. Global warming is still an issue where there are many skeptics who do not believe that this phenomena is occurring.

Crisis as Slow-Moving: Heat Waves

Heatwaves in Europe: A Deadly Combination

The case study of this chapter focuses on heat waves in Europe. While severe heat is survivable, many areas of the world are poorly equipped for handling heatwaves due to a lack of cooling systems (i.e., air conditioners). In 2003, Europe experienced its hottest summer on record since 1540 resulting in the deaths of more than 70,000 people and the loss of several food crops (wheat and grapes in particular) due to drought. In 2006, Europe experienced another extreme heat wave mostly affecting Western Europe including the western parts of Russia. In 2007, another severe heat wave affected Europe, this time mostly centered in southern and Eastern Europe. This is not to imply that Europe is the only area of the globe affected by heat waves as they can occur in any part of the world.

According to the EuroHEAT project, an initiative of the World Health Organization (WHO) Regional Office of Europe, "there is no standard definition for a heat wave" (European Commission's Directorate for Public Health and Risk Assessment). Robinson (2001) points out regarding defining heatwaves:

> A basic definition of a heat wave implies that it is an extended period of unusually high atmosphere-related heat stress, which causes temporary modifications in lifestyle and which may have adverse health consequences for the affected population. Thus, although a heat wave is a meteorological event, it cannot be assessed without reference to human impacts.

Heatwaves are characterized by warm temperatures that last over a longer than expected period of time. However, in line with EuroHEAT, Robinson's explanation highlights how defining a weather phenomenon as a heatwave is interconnected with the impact it has on people.

Medically, deaths due to heat waves are caused by hyperthermia when the body temperature rises too much (the opposite of hypothermia). The elderly and those with chronic medical conditions tend to be more susceptible to hyperthermia (Pappas, 2016). While this is in part due to individual physical condition and health, medical historian and bioethicists Richard Keller explains that this trend is also due to social isolation; for example:

> In France in 2003, the heat hit in August, when many Europeans go on vacation. Elderly people found themselves in mostly empty apartment buildings when

> the heat crisis reached them. Some were found dead with their doors ajar, Keller said, suggesting that they were trying to get out and get help when they collapsed. (Pappas, 2016)

Keller also points that some may die in a heatwave due to a lack of awareness of the crisis state they were in: "some people during the heat wave probably thought they were going through an uncomfortable time and didn't recognize how precarious their survival was." Emphasizing this point, Klinenberg (2002) argued that in comparison to the dramatic visibility of other extreme weather events such as hurricanes, tornadoes, earthquakes, and floods, extreme heat is "invisible."

Heatwaves as a crisis have some similarities with the winter ice storm disaster in Canada described previously. Some areas may be better prepared to handle them than other areas. For instance, while an area such as Chicago in the U.S. midwest may be poorly equipped to handle a heat wave due to the lack of air conditioning in older buildings and residential units occupied by those with limited economic resources, a place such as Arizona located in the desert southwest possesses the infrastructure to handle an extreme weather condition such as a heatwave.

Summary

The example of heatwaves illustrates a creeping crisis and the problems associated with problem identification. Heat is to be expected during the summertime, however, heatwaves can be deadline, especially to vulnerable groups such as the elderly. As a crisis, a heatwave is difficult to define, making the determination of a response all the more difficult. Even though heatwaves are slow-moving, they are now a familiar aspect of the summer European climate. As explained by Public Health England:

> The climate is changing and current analysis in the UK Climate Change Risk Assessment 2012 suggests that summers are going to get hotter in the future: "despite the uncertainties related to future climate change and its impacts, the evidence is now sufficient to identify a range of possible outcomes that can inform adaptation policies and planning. …" (p. 10)

The purpose of this chapter was to provide an overview of crisis communication as a field of study and to describe the relationship between crisis communication and sensemaking. A key emphasis of this chapter was on how problems themselves are communicatively constituted. A given circumstance may be a problem for one group and not for another group. This situation is related in part to material circumstances and technology in terms of how one group may be better prepared to adapt to extreme weather circumstances than another group. In addition, pacing or timing plays a role in recognizing a circumstance as problematic such that a response needs to occur. With slower moving situations, it may be difficult to recognize the tipping point from discomfort and inconvenience to medical, organizational, or societal emergency. However, the example of European heatwaves illustrates how slow-moving does not have to equate to being unidentifiable. While there is uncertainty in predicting exact climate circumstances, there is enough information to anticipate future trends and crises.

Note

1. Weick (1995) borrowed this phrasing from social psychologist Graham Wallas (1926).

References

Bergeron, C. D., & Cooren, F. (2012). The collective framing of crisis management: A ventriloqual analysis of emergency operations centres. *Journal of Contingencies and Crisis Management*, *20*(3), 120–137. https://doi.org/10.1111/j.1468-5973.2012.00671.x

Buttny, R. (2004). *Talking problems: Studies of discursive construction*. Albany, NY: State University of New York.

Castor, T., & Bartesaghi, M. (2016). Metacommunication during disaster response: 'Reporting' and the constitution of problems in Hurricane Katrina teleconferences. *Management Communication Quarterly*, *30*, 472–502.

Castor, T., & Cooren, F. (2006). Organizations as hybrid forms of life: The implications of the selection of human and non-human agents in problem-formulation. *Management Communication Quarterly*, *19*, 570–600.

Czarniawska, B. (2009). Introduction: Action nets. In B. Czarniawska (Ed.), *Organizing in the face of risk and threat* (pp. 1–8). Glasgow: Edward Elgar.

Fairhurst, G. T., & Cooren, F. (2009). Leadership as the hybrid production of presence(s). *Leadership*, *5*(4), 469–490. https://doi.org/10.1177/1742715009343033

Fearn-Banks, K. (2011). *Crisis communications: A casebook approach* (4th ed.). New York, NY: Routledge.

European Commission's Directorate for Public Health and Risk Assessment. (n.d.). *Public health responses to heat waves*. Retrieved January 11, 2017 from http://ec.europa.eu/health/climate_change/extreme_weather/heatwaves_en#fragment1

Gephart, R. P. (2007). Crisis sensemaking and the public inquiry. In C. M. Pearson, C. Roux-Dufort, & J. A. Clair (Eds.), *International handbook of organizational crisis management* (pp. 123–160). Los Angeles, CA: Sage.

Grint, K. (2005). Problems, problems, problems: The social construction of 'leadership'. *Human Relations*, 58(11), 1467–1494. Retrieved March 5, 2013 from http://hum.sagepub.com/cgi/doi/10.1177/0018726705061314

Klinenberg, E. (2002, July 30). Dead heat: Why don't Americans sweat over heat-wave deaths? *Slate*. Retrieved January 11, 2017 from http://www.slate.com/articles/health_and_science/medical_examiner/2002/07/dead_heat

Maitlis, S., & Christianson, M. (2014). Sensemaking in organizations: Taking stock and moving forward. *The Academy of Management Annals*, 8(1), 57–125. https://doi.org/10.1080/19416520.2014.873177

Maitlis, S., & Sonenshein, S. (2010). Sensemaking in crisis and change: Inspiration and insights from Weick (1988). *Journal of Management Studies*, *47*(3), 551–580. https://doi.org/10.1111/j.1467-6486.2010.00908.x

McConnell, A. (2003). Overview: Crisis management, influences, responses and evaluation. *Parliamentary Affairs*, 56(3), 363–409.

Murphy, R. (2004). Disaster or sustainability: The dance of human agents with nature's actants. *Canadian Review of Sociology & Anthropology*, *41*(3), 249–266.

Pappas, S. (2016, June 20). Deadly degrees: Why heat waves kill so quickly. *Live Science*. Retrieved January 11, 2017 from http://www.livescience.com/55129-how-heat-waves-kill-so-quickly.html

Public Health England. (2015). *Heatwave plan for England*. London: Public Health England.

Rittel, H. W. J., & Webber, M. M. (1973). Dilemmas in a general theory of planning. *Policy Sciences*, *4*, 155–169.

Robinson, P. J. (2001). On the definition of a heat wave. *Journal of Applied Meteorology*, *40*, 762–775.

Seeger, M. W., Sellnow, T. L., & Ulmer, R. R. (1998). Communication, organization, and crisis. *Communication Yearbook*, *21*(23), 1–276.

Wallas, G. (1926). *The art of thought*. New York, NY: Harcourt Brace.

Weick, K. E. (1988). Enacted sensemaking in crisis situations. *Journal of Management Studies*, *25*, 305–317. https://doi.org/10.1111/j.1467-6486.1988.tb00039.x

Weick, K. E. (1995). *Sensemaking in organizations*. Thousand Oaks, CA: Sage.

Weick, K. E. (2001). *Making sense of the organization*. Malden, MA: Blackwell.

Weick, K. E., Sutcliffe, K. M., & Obstfeld, D. (2005). Organizing and the process of sensemaking. *Organization Science*, *16*(4), 409–421. https://doi.org/10.1287/orsc.1050.0133

· 4 ·

MAKING SENSE OF A WINTER STORM

In January 28–29, 2014, the southeast United States experienced an unusual and severe winter snow storm, named Leon, that left inches of snow and sub-freezing temperatures in a swath of land that spanned from Texas to Virginia. During this unusual weather event, the state of Georgia was especially unprepared and experienced traffic jams, saw abandoned vehicles on the freeway and children left at their schools overnight due to parents being unable to make the drive to pick them up (Carter, 2014). This lack of preparedness on the part of Georgia made headline news with the leadership of Georgia being ridiculed and the governor in turn issuing apologies (Breslin, 2014). Subsequently, the governor formed a task force to investigate the problems associated with the winter storm response with the task force developing a revised plan for handling future extreme winter weather situations ("Governor's Severe Winter Weather Warning and Preparedness Task Force Report", 2014).

All three main CCO approaches address the issue of how organizations are constituted over *time*: the Montreal School is attuned to how the "there and then" is brought into the "here and now" in organizational conversations (Cooren, 2004); the Four Flows approach highlights "reflexive self-structuring" (McPhee & Zaug, 2000); and, Luhmann's systems approach addresses autopoetic self-production (Luhmann, 1982). Each of these characterizations points to the processual nature of constitutive activity, reflecting Weick's recommendation that we think of organizations in terms of processes rather than products, verbs (organizing) rather than nouns (organization).

In this chapter, I focus on sensemaking. While sensemaking studies have addressed how people have *retroactively* made sense of a crisis, another way that sensemaking can be applied is to make sense of a problem *before* it occurs in anticipatory or prospective sensemaking. This is a challenging activity to do, but an important skill to develop given that if you can see that a problem is unfolding, you can be empowered to make choices that could prevent the problem, minimize its consequences, and if needed, be quickly prepared for recovering from the crisis. A key concept that I introduce as a way of reflecting on a potentially problematic situation is the idea of *responsiveness*, borrowed from Russian literary critic, Mikhail Bakhtin (1981).

In this chapter, I analyze Georgia leadership's response to winter storm Leon as a case study for illustrating the role of time and *responsiveness* in constitutive processes. Georgia's leadership failed to make sense of the storm as a *potential crisis* as it was approaching. This analysis is intended to contribute to the understanding of the communicative constitution of organizations by advocating the inclusion of time and timing as a significant feature for understanding the communicative constitution of organizations. The future is already a part of many significant organizational communication activities such as strategic planning, creating vision statements, goals, and policy documents. With the specific area of crises, organizations at once are addressing past and future projections.

This chapter is also relevant for understanding leadership sensemaking processes with respect to climate and weather crises. While climate change is reaching the point where it has become broadly accepted with key public disputes over the causes rather than existence of climate change, Lorenzoni and Pidgeon (2006) argued that climate change should be viewed from the perspective of local experiences. The case study illustrates how local experiences can shape how one responds to an emergency.

Organizational Sensemaking

Sensemaking is a term that has been used by several scholars who have theorized about this concept in different ways. Weick (1995) provided a very comprehensive overview of sensemaking as a concept as well as a thorough summary of key findings from organizational sensemaking studies. In communication, sensemaking has also been popularized by the research of Brenda Dervin (see Dervin, Foreman-Wernet, & Lauterbach, 2003). Dervin, coming from the area of information science, has made important contributions to the understanding of sensemaking as a cognitive process. This chapter focuses on Weick's contribution to sensemaking given that his theorizing and research specifically address organizational sensemaking. The two key areas I highlight in this overview are the process of sensemaking and the seven characteristics of organizational sensemaking.

The Process of Sensemaking

Sensemaking has been categorized as a systems approach to organizational communication (e.g., Eisenberg, Goodall, & Trethewey, 2013; Mumby, 2012). A *system* is a unit of entity that consists of interdependent parts. Initially developed as a biological concept, the idea of a system also highlights the processes that are involved with maintaining a system. For living creatures, this would involve taking in food and water, among many other processes. For organizations, this could involve making and selling products, among many other activities. In viewing sensemaking as a part of a systemic process, Weick (1979) described three phases of the organizational sensemaking process: enactment, selection, and retention.

Each of the following quotations are taken from Weick's (1995) explanation of sensemaking:

> How can I know what I think till I see what I say? (p. 12)
> How can I know what I value until I see where I walk? (p. 183)
> How can I know what we did until I see what we produced? (p. 227)

The preceding quotations get at the process as well as complexity of sensemaking. While sensemaking is about how one gives meaning to experience, Weick has highlighted that sensemaking is difficult to do when one is in the middle of a situation. Rather, sensemaking is something that often occurs after the fact when one is able to reflect back upon a situation. As an example, consider the 9–11 terrorist attacks in New York and Washington D.C. I vividly remember that morning as I heard news reports of planes crashing in New York city. As this event was happening, many were not sure what was going on, and in the days following, sensemaking of the event as terrorist attack occurred.

Weick (1979) outlined three phases of organizational sensemaking: *enactment*, *selection*, and *retention*. Enactment refers to when organizational members attend to "cues" and realize that something is occurring. However, Weick has emphasized that a key aspect of enactment is not that we realize that something is going on *but* that we construct our situations such that we determine something is occurring. In other words, enactment is not merely reacting, but constructing; Weick (2001) explains:

> *Enactment* is the social process by which a "material and symbolic record of action" (Smircich and Stubbart, 1985, p. 726) is laid down. The process occurs in two steps. First, portions of the field of experience are bracketed and singled out for closer attention on the basis of preconception. Second, people act within the context of these bracketed elements, under the guidance of preconceptions, and often shape these elements in the direction of preconceptions (Powers, 1973). Thus, action tends to confirm preconceptions. (p. 226)

Weick describes enactment as a form of bracketing or of attending to some cues over others. In this respect, in enactment we socially construct our environments.

A key point of Weick's regarding enactment is that the process of enactment constructs environments or the conditions and circumstances of an organization. Moreover, crises are enacted, meaning, that crises themselves are socially constructed. As Weick (2001) explains:

> Thus, from the perspective of enactment, what is striking is that crises can have small, volitional beginnings in human action. Small events are carried forward, cumulate with other events, and over time systematically construct an environment that is a rare combination of unexpected simultaneous failures. (p. 228)

As an example for this, Weick analyzed the Union Carbide Bhopal plant toxic gas leaked that occurred in India in 1984. This tragedy, which cost the lives of 3,787 people, was caused in part by a leaky gas pipe. Weick was critical of analyses that pointed to the gas leakage as the cause of the tragedy and instead, directed attention to flaws in a pipe cleaning process when a "slip bind" was not reinserted which then "allowed water to back up and enter the MIC tank and catalyze a complex chemical interaction" (p. 228). The key point Weick made here is that there is a relationship between crises and human actions (or inactions).

The second phase of the sensemaking process is selection which refers to interpretations or giving meaning to data and sensory experience. Weick argued that organizations are interpretation systems. The third phase of sensemaking is retention which is when organizational members make decisions on what they have learned through the enactment and selection phases and decide what processes and procedures to retain to apply in future circumstances.

There are a number of implications of organizational sensemaking which Weick (1995) summarized as seven properties of sensemaking:

1. Grounded in identity construction
2. Retrospective
3. Enactive of sensible environments
4. Social
5. Ongoing
6. Focused on and by extracted cues
7. Driven by plausibility rather than accuracy (p. 17)

There are several compelling points that Weick makes regarding the process of organizing. Also, Weick's attention to interpretation, identity construction, social construction, and the social nature of sensemaking points to a great deal of compatibility along with differences between Weick's sensemaking approach and CCO perspectives.

CCO and Sensemaking

While Weick's approach has made important contributions to the advancement of organizational communication generally and to interpretive and constitutive approaches specifically, there are some differences with CCO approaches. Taylor and Van Every (2000) issued a very extensive critique of Weick's approach. One key point that they made is that while Weick recognizes the socially constructed nature of organizations, his work does not quite address how *organizations* are themselves socially constructed, or more specifically, constituted. Rather, his work has explicated how environments and crises are constructed rather than how the organization itself is constructed.

A second critique is that Weick's work does not provide enough depth in explaining communication processes (also see Castor & Cooren, 2006). In other words, while Weick argues that organizations are social interpretation systems, how interpretations occur within communication and social interaction are not described or explained.

Although there are differences between CCO and Weick's approach, Weick's sensemaking perspective has provided an important foundation for interpretive and constitutive approaches to organizing. Weick's sensemaking approach is especially relevant for this project given the extensive work he has completed in analyzing crises and how crises are enacted. In the remainder of this chapter, I focus on the issue of time in extending Weick's sensemaking approach. I also emphasize a focus on language-use during pre-crisis organizational communication as a way of addressing one CCO critique of the limitation of Weick's work in providing communication-oriented explanations of sensemaking processes.

The Relevance of Time in Constitutive Processes

Time and timing are important concerns for organizations and in organizational communication. Organizations develop mission and vision statements to guide actions toward a particular future; goals are established as objectives for the future; policy and procedure manuals are developed to guide future actions. Time is also a crucial aspect of theorizing in CCO approaches, however, connections to the past tend to be emphasized.

Putnam and Nicotera (2009) described that CCO approaches examine processes of organizing as they "endure across time and space" (p. 4). "In addition, CCO approaches move beyond local constructions to illustrate how dislocal practices and resources draw on both current and past actions and structures" (Putnam & Nicotera, p. 5). In the Montreal school, Benoit-Barnat and Cooren (2009) describe "*presentification*" and how the past of the organization (the "there and then") is brought into the present (the "here and now"). Cooren and Fairhurst (2004) address "spacing" and "timing" and "*spatio-temporal closures*, which define the structures of organizing processes" (pp. 794–795). Their analysis included a consideration of the role of human and nonhuman agency in organizational closure. They introduce and apply the concept of *schema* to describe how organizational closures are achieved in interactions. They also acknowledge that "each initiation of a schema opens a specific space and a specific time whose closure can be *anticipated* and *projected* (but never determined)" (p. 814).

The Four Flows model evokes time and process as organizations are constituted by flows. The Social Systems perspective emphasizes the centrality of decisions in organizations as operations fundamentally involved with the reduction of risk and uncertainty, concepts that connect the past and future. Czarniawska (2013) called for a re-focusing of attention from actor-networks to activity networks: a shift from things/objects/nouns to activity/doing/"verbing." Processes and activities are fundamentally connected with time. CCO approaches however have typically dealt with how the past is brought into the present, rather than how the future is evoked in the present.

It is also through time that connections are made between materiality and constitutive processes:

> One approach to materiality in CCO theories is the actor-network notion of inscription; that is, the material and structural exist as texts, such as documents and policies reified from past interactions, and as material objectives, such as buildings and furniture, that impose on and are shaped by organizing processes. (Putnam & Nicotera, 2009, p. 163)

Material things such as policy documents and buildings function as objects developed in the past that in turn shape the present. However, this relationship can be extended to discussing how the future can impact the present when considering or anticipating forthcoming events and developments.

Although time and timing have been addressed in CCO approaches, CCO approaches have tended toward a bias of past and present orientations in looking at how present interactions build upon and incorporate past texts, conversations, patterns and/or rules. However, organizations are called upon to connect the present to the future on a regular basis when creating vision statements, strategic plans, or projection models. Moreover, failure to connect the present to the future can result in the failure of the organization. In addition, when CCO oriented studies have examined problems, the focus has been on how participants make sense of problems after the fact or retrospectively. In the following section, I advance the notion of "responsiveness" as a component to consider in the communicative constitution of organizations in understanding one way that organizational members can use their communication practices to project into the future.

Responsiveness

The notion of responsiveness is based on the work of Russian literary critic Mikhail Bakhtin (1981) and communication scholar John Shotter (1993) and is closely related to the communication practice of dialogue. Bakhtin's notion of dialogue was initially based on examining language in novels, however, his theorizing has been applied in many other contexts such as interpersonal relationships (see Baxter & Montgomery, 1996) and organizational studies (Beech, 2008; Deetz & Simpson, 2004; also see Anderson, Baxter, & Cissna, 2004).

There are two ways that dialogue and responsiveness are related. In one sense, responsiveness may be considered from within interaction as responding to the immediate situation or what another has said (McNamee & Shotter, 2004). A second sense of responsiveness is anticipatory. As Bakhtin (1986) explained:

> All real and integral understanding is actively responsive, and constitutes nothing other than the initial preparatory stage of a response (in whatever form it may be actualized). *And the speaker himself [sic] is oriented precisely toward such an actively responsive understanding*. ... [The speaker] expects response, agreement, sympathy, objection, execution, and so forth. (emphasis added, p. 69)

In other words, when speakers speak (or communicators communicate), it is with the expectation and anticipation of future responses from others: "from the very beginning, the utterance is constructed while taking into account possible responsive reactions, for whose sake, in essence, it is actually created" (p. 94). There is an anticipated audience with anticipated reactions to whom one is speaking and directing one's utterances toward. Granted, at the individual and group level people vary in their anticipatory skills. However, it is through the notion of responsiveness as anticipatory that we can discuss and describe how the future may be brought into the presence of a given interaction.

Shotter (1993) adds to Bakhtin's notion of responsiveness by introducing the notion of rhetorical responsiveness (also see Cunliffe, 2002; Shotter, 2004, 2006). Social actors cannot respond in any way that they see fit, but must act from within the constraints of a given situation to show that they are socially accountable. Shotter, along with Schön (1983) focused on the "messiness" of communication when "acting from within" a given situation. In contrast to post-hoc analysis where one has the luxury of time, perspective, and outcomes to assess a situation and describe what happened, Shotter, drawing from Wittgenstein, highlights that during decision-making, one is acting from within a given situation, trying to make sense of ambiguous and as yet undetermined circumstances. In other words, Shotter, like Weick, recognized that it is difficult to figure out situations as they are happening.

Schön's description of the "reflective practitioner" is relevant here. Schön emphasized the importance of reflecting upon one's actions and learning from the situations one is in and experiences one has. This emphasis is significant when contrasted with other ways that learning could occur such as through reading descriptions and theory. While reading obviously is important, it is a different type of knowledge as compared to learning from experience. As some of my students would likely paraphrase, it is about being "street smart."

Shotter and Schön, in applying their analysis to the challenges of organizational management, highlighted how there are different perspectives on management and knowing. There is "*knowing about*" or possessing theoretical knowledge. There is "*knowing how*" which is technical knowledge of how to do something. Finally, there is "*knowing from within*" which can be described as "practical knowledge." This is knowing how to act from within a given situation, to respond in the moment to unfolding circumstances. Shotter has described this way of knowing and acting in relation to management as "practical authorship" (see Holman & Thorpe, 2002) in that the manager is not just responding or reacting, but "authoring" or creating circumstances.

Shotter's description of practical authorship and Schön's reflective practitioner can be compared with Weick's (2001) description of heedfulness: "in heedful performance, each action is modified by its predecessor (Ryle, 1949, 1949, p. 42). In heedful performance, the agent is still learning" (p. 264). Relatedly, Weick's notion of improvisation in organizations points to the need to respond in innovative ways to organizational circumstances.

The case of Georgia's lack of preparedness for winter storm Leon in January, 2014, is a case where organizational leaders failed to improvise in a heedful way. They were not able to successfully "act from within" the circumstances of the winter storm. In highlighting responsiveness, my purpose is to explicate this failure to "act from within" as a discursive process where material circumstances are intertwined with constitutive processes. In other words, constitutive activities do not just consist of bringing the past into the present, but also consist of bringing a projected or anticipated future into the present. As Bakhtin's (1981) description of dialogue pointed out, when we communicate, there is a future orientation in that we communicate with an anticipated audience in mind who will have an anticipated response. Organizations are not just grounded in the past and patterns from the past. They are also constituted through communication that anticipates a future and that is responsive to an anticipated audience and conversation.

Constituting Winter Storm Leon as Disaster

During the winter of 2014, there were several extreme winter weather emergencies across the globe due to a change in what is known as the polar vortex, a pocket of very cold air that is present over the polar region during winter months (MacMath, 2014). In 2014, there was a shift in the pattern of the polar vortex that allowed for the cold air over the polar region to flow into parts of North America, resulting in extremely cold temperatures in some areas and winter snow storms in parts of the country that did not typically experience such weather.

In late January, the National Weather Service (NWS) predicted a winter storm that would go through a southeastern swath of the United States, affecting the southern parts of Texas and Louisiana and larger areas of Georgia. A *Cable News Network (CNN)* report outlined the timeline for the NWS forecasting as follows. On January 26 (Sunday), the National Weather Service (NWS) issued their winter storm watch for Georgia, and the following day (4:53 a.m., Monday), they added the metro Atlanta area to the winter storm watch with a predicted snowfall of up to two inches (Wilson & Smith, 2014). Later that day (3:22 p.m.), the NWS updated their warning to a watch. By Tuesday (3:38 a.m.), "Atlanta and its surburbs are placed under a winter storm warning effective at 9 a.m." (Wilson & Smith, 2014). However, by Tuesday at 8:00 a.m., commuters began their morning drive to work. In a prescient statement, one blogger predicted: "Southern drivers, unaccustomed to such conditions, will be slippin' and slidin', and power outages are likely where the precipitation falls as freezing rain" (Plotkin, 2014).

The state of Georgia in general and the city of Atlanta in particular were unprepared for the one to two inches of snow and freezing that occurred. Motorists were stranded on freeways as there were more and more traffic jams. Powerlines were downed due to ice freezing on them. Many parents were unable to reach their children at school, and the children subsequently spent the night at their schools. The leadership of Georgia was criticized and ridiculed within the media. Georgia Governor Nathan Deal responded in the media with an apology and acceptance of responsibility (Winograd, 2014). Governor Deal immediately formed a Severe Winter Weather Warning and Preparedness Task Force that was charged with investigating the specific problems associated with Georgia's storm Leon response and providing recommendations for the future (see "Severe Winter Weather Warning and Preparedness Task Force").

For my discursive units of analysis, I first address weather forecasting as pragmatic speech act. I draw from a key theory that has influenced the Montreal School approach, *Speech Act Theory* (Searle, 1969) to analyze weather forecasting in terms of its implications for responsiveness. In examining how

Georgia leadership responded to the crisis, I utilize *accounts* as a sensitizing analytic tool. Accounts are linguistic devices offered during a social and/or moral breach (Scott & Lyman, 1968). Accounts are typically offered after the fact as a way of reconciling the account giver and restoring his or her standing in the community. Accounts have also been analyzed as everyday communication that makes actions intelligible and accountable (for thorough overviews see Buttny & Morris, 2001).

Weather Forecasting as Pragmatic Speech Act

In National Weather Service language, a "watch" is when significant winter weather is expected with 12–36 hours of advance notice. A "warning" is when a severe storm is likely to pose a hazard and threat to life to property. With a warning, the storm is likely, occurring, or imminent.

More formally, to draw from Speech Act Theory, warnings fall under the category of directive (Bach, n.d.) which is a speech act that is intended to commit the hearer to a particular course of action. In the case of weather forecasting, warnings are about committing weather phenomenon to a course of action. However, there are a number of ambiguities here. First, weather forecasting may be construed as *descriptive* describing a state of affairs. Weather forecasting may also be construed as warning or as being akin to a threat or promise. In the case of weather forecasting of extreme weather events, with a warning, the speaker (meteorologist) is speaking on behalf of a weather event and the negative outcomes that are predicted to come from that weather event in the case of extreme weather phenomena.

Weather forecasting is descriptive, but when the topic of the forecast is a severe weather event, forecasting is *action-implicative* in that actions are implied rather than specified. Another way to conceptualize this is that forecasting may indicate danger, but not specify problems and solutions. Within a network of agencies and action, forecasting points to an immediate chain of danger, but meteorologists may not go further down the chain to tell us or forecast traffic jams. Those rather are problems that lie within the realm and jurisdiction of others to enact and/or foretell. It is up to public officials and emergency managers to translate forecasts of dangers to dangers of snowy roads and icy power lines.

According to Georgia's Severe Weather task force report, the language use of warning and watch were ambiguous. In the report section on National Weather Service forecasting, the task force highlighted that "the public must have access to weather information that they can both understand and use when making decisions" (p. 10) and that "the first line of defense against a weather emergency is foreknowledge" (p. 10). These statements imply that accurate weather information was not accessible or understandable by the public. This latter point is especially emphasized by the second recommendation regarding weather forecasting provided by the task force: "create uniform definitions and criteria for severe weather terminology" (p. 11). This recommendation focused on the interpretation of communication rather than on the nature of the information. In locating the problem as a lack of clarity in terminology, the task force placed blame on the weather forecasting terminology:

> currently, there is inconsistency in the criteria (weather conditions) used to determine when to employ specific severe weather terms such as *Watch*, *Advisory*, and *Warning*. This creates confusion and uncertainty, both for the public in general and for decision-makers. … (p. 11)

The reactions to the aforementioned labels can also be seen in the corresponding timeline of how Georgia leadership reacted, basically waiting until the "watch" was upgraded to a "warning." This analysis is not intended to cast blame upon NWS terminology, but rather to present part of the context for the timing of when Georgia leadership reacted to the storm. Also, I wish to note that the National Weather Service does provide descriptions for watch, advisory, and warning. Therefore, confusion seemed to lay more with interpreting those terms and in understanding the public service decisions that they imply.

Moving From Storm to Disaster

As previously described, Georgia Governor Deal's immediate public response was to apologize to the public, accepting personal responsibility for the state's inadequate response, stating "I accept responsibility for the fact that we did not make preparation early enough to avoid these consequences … I'm not looking for a scapegoat. I'm the governor, the buck stops with me" (Botelho & Watkins, 2014). While Deal accepted responsibility and took active and visible actions to prevent future similar disasters, there were other ways of framing the problem and its causes that were implicit in these actions and statements.

In one statement, Deal indicated that his "in the moment" response was contingent upon the Georgia Emergency Management Agency response. In an email, one state employee stated on Monday, 3 p.m. (approximately a day before the full brunt of Leon): "everyone keeps telling me how bad the weather is going to be and I keep saying if the weather was going to be bad, Charley would have called and he hasn't called me" ("Emails show casual response ..."). In this rich and telling email, the state employee indicated awareness of the potential dangers of the weather, however, there is another agent in the system that the employee orients to more strongly and that is the Georgia Emergency Manager, Charles English. However, GEMA head Charles English was criticized for showing a "casual" attitude toward the coming winter storm.

As English was receiving inquiries and questions, his responses showed that he did not take the storm seriously. In one email exchange that occurred two days before the storm, English responded to concern by stating, "I heard some broadcast meteorologists predicting something" ("Emails show casual response ..."). A day before the storm, a GEMA employee asked English if he should cancel a planned trip to Washington D.C., to which English responded, "Thank you. Sure is warm outside" ("Emails show casual response ...").

By Monday afternoon, television news channels were forewarning their viewers that the storm was going to be severe. While some within the Georgia state administration were expressing concern and asking about what should be done in anticipation of the storm, a key decision-maker, GEMA manager Charlie English, was anticipating something more moderate than what actually occurred despite meteorological warnings.

To draw from a similar work, in their study of Governors Blanco and Schwarzeneggar communication during state emergencies, Fairhurst and Cooren (2009) argued that one difference in the communication of the governors was how they "mobilized" nonhuman actants through their discourse with Blanco failing to do this and Schwarneggar successfully doing so. This analysis of Governor Deal's response adds an additional layer to how mobilization occurs in that time and timing are crucial elements in this process. The appropriate actants were eventually mobilized or called into action in the form of trucks to treat the snowy and icy roads, but this happened at a time that was too late to be effective. In addition to being too late, there was a crucial actant that was ignored until too late and that was the storm itself.

As the storm was unfolding, it did not present a danger, but it was at its later state when it would be dangerous as the inches of snow fell onto the unsalted roads. At that point, it was too late to prepare. Actions would have been *reactive* when, it order to be effective, actions needed to be *proactive*. In other words, actions needed to be *responsive* to a projected future. Just as Bakhtin (1991) argued that dialogue is expressed in response to an anticipated audience, Georgia's crisis response needed to occur in anticipation of the storm. The storm was not taken seriously as an actant in the upcoming course of weather events in Georgia.

The report of the extreme weather task force itself is indicative of a network of agents and activity that is activated or failed to be activated in the course of weather danger. The report itself highlighted several recommendations that included as a primary recommendation placing ultimate authority for decision-making into the hands of the Georgia Emergency Management Agency head. The recommendation implies that this wasn't done initially and that there was confusion over who was in charge—this despite earlier emails that showed state employees looking to the GEMA head for a decision and the GEMA head providing responses that indicated an awareness of the snowstorm and a decision that it was not going to be a serious problem. The decisions to respond were made as incidents were occurring

Coda

Georgia leadership did have an opportunity for a "do-over" with regards to their winter storm response (McCaffrey, 2014). Two weeks after winter storm Leon incapacitated Atlanta, another severe winter storm came through the Atlanta area. This time, in advance of the winter storm, Gov. Deal "declared a weather-related state of emergency for 45 counties" (Payne, 2014, para. 5). Atlanta Public Schools preemptively closed schools for two days. This time, Georgia and the city of Atlanta were prepared, thus averting a crisis in the face of another winter storm.

This do-over highlights the importance of prior experience in effective crisis response. Conversely, it also highlights the challenge in knowing what to do in the face of a novel crisis. In addition, while many cities and states do know what to do during a predicted winter storm, such a storm was a new experience for the leadership and residents of Georgia, illustrating the significance of one's local experiences in shaping how events are interpreted and subsequently, how responses are determined.

Summary

While Winter Storm Leon was severe, its immediate impact of 1–2 inches of snow is relatively minor as far as winter weather is concerned. However, this impact must be considered within context. In areas that are used to severe winter weather, 1–2 inches is minor, whereas for areas that rarely experience snow, even 1–2 inches is a major event, especially when coupled with ice. One way to view this event is that because it was so unusual compared to typical Georgia weather, there were no prior routines or repertoires established for how to interpret and respond to severe winter weather. However, given the challenges to typical weather patterns from climate change, old weather patterns can no longer be relied upon as predictors of new weather patterns.

The response of Georgia to winter storm Leon illustrates points made by Beck (1992) and Murphy (2004) regarding the relationship between risk and technology. They argued that risk from climate, environment, etc. do not just "present themselves" but rather, they are constructed through our very technologies that are designed to make our lives better. With the case of Winter Storm Leon, this weather event was constructed as a crisis due to the failure to anticipate and prepare. To emphasize, the storm itself in terms of snow, cold, rate of fall, etc. were not socially constructed. However, what was constructed was the interpretation of the meaning of the storm. In Georgia leadership's initial construction of the storm, it was not anything to be concerned about. However, this construction did not mesh with the actual physical materiality of the storm. As a consequence, the crisis was "enacted" through inactions such as not calling out salt trucks or closing schools.

CCO approaches have in common the assumption that communication is constitutive of organizing. They call upon assumptions common to a system's approach vis-à-vis Karl Weick's work and as such, have tended to focus on patterns enacted within a system and how past actions shape and are brought into the present. The phrasing of "*translocal*" also indicates how CCO approaches have taken into consideration how communication has functioned to bridge spatial distance and separation. Through this chapter, I illustrated how the bridging of time is also crucial in the communicative constitution of organizing. While communication scholars have illustrated how the past is called into the present, there are also ways that the future is called into (or not) the present. The problems associated with the Georgia leadership response to winter storm Leon may be described as a failure to project down the chain or network of agencies, or to project appropriately into the future. They weren't being responsive to the appropriate actants. In the case of winter storm Leon, public officials were *reactive* rather than *responsive*. They reacted in the moment as snow was accumulating, but this was too late to send out the road treatment trucks to treat the roads (especially as rush hour commute was starting). A *responsive* approach would react both to the past *and* to an anticipated future.

Conclusion

Through this chapter, the key issue that I addressed was Weick's process of sensemaking and how to strengthen its connection to CCO concerns regarding communication. The key case study consisted of Georgia's winter storm crisis. Tracing back through the communication showed an awareness of the potential dangers from the storm, but also a failure to act upon the forecasts. One way to interpret this failure to act is as a failure to anticipate the network of agents and activities involved with severe weather. Moreover, the network must be viewed not just in spatial and translocal terms, but also in terms of time, activity and process. The latter two terms are ones that are familiar to organizational communication scholars as being inherited from a systems theory approach. However, the time orientation inherited from systems theory is past and present-oriented. One unique feature of humans is the use of language, which also involves the ability to imagine, including imagining the future. When constituting organizations and crises, we are not confined to bringing the past into the present but also have the option to bring an anticipated future into the present. *How* that future is constructed varies and is shaped by present actions.

References

Anderson, R., Baxter, L. A., & Cissna, K. N. (Eds.). (2004). *Dialogue: Theorizing difference in communication studies*. Thousand Oaks, CA: Sage.

Bach, K. (n.d.). Speech acts. *Routledge encyclopedia of philosophy*. Retrieved from http://online.sfsu.edu/kbach/spchacts.html

Bakhtin, M. M. (1981). *The dialogic imagination*. Austin, TX: University of Texas.

Bakhtin, M. M. (1986). *Speech genres and other late essays*. C. Emerson & M. Holquist (Ed.). (V. W. McGee, Trans.). Austin, TX: University of Texas.

Baxter, L. A., & Montgomery, B. M. (1996). *Relating: Dialogues and dialectics*. New York, NY: Guilford.

Beck, U. (1992). *Risk society: Towards a new modernity*. Los Angeles, CA: Sage.

Beech, N. (2008). On the nature of dialogic identity work. *Organization, 15*(1), 51.

Benoit-Barnat, C., & Cooren, F. (2009). The accomplishment of authority through presentification: How authority is distributed among and negotiated by organizational members. *Management Communication Quarterly, 23*(1), 5–31.

Botehlo, G., & Watkins, T. (2014, January 30). Georgia officials under fire for actions before, during, after snow. *CNN report*. Retrieved from http://www.cnn.com/2014/01/30/us/winter-weather/

Breslin, S. (2014, February 12). Georgia Governor Nathan Deal's staff acknowledges dangers of winter storm Leon 24 hours in advance; apologies for Atlanta gridlock. Retrieved from http://www.weather.com/news/commuter-conditions/winter-storm-leon-causes-massive-gridlock-atlanta-20140128

Buttny, R., & Morris, G. H. (2001). Accounts. In W. P. Robinson & H. Giles (Eds.), *The new handbook of language and social psychology* (pp. 285–301). New York, NY: Wiley.

Carter, C. J. (2014). Ice storm wallops southeast, stranding drivers, cutting power. Retrieved from http://www.cnn.com/2014/02/12/us/winter-weather/

Castor, T., & Cooren, F. (2006). Organizations as hybrid forms of life: The implications of the selection of agency in problem formulation. *Management Communication Quarterly*, *19*(4), 570–600.

Cooren, F. (2004). The communicative achievement of collective minding: Analysis of board meeting excerpts. *Management Communication Quarterly*, *17*(4), 517–551.

Cooren, F., & Fairhurst, G. T. (2004). Speech timing and spacing: The phenomenon of organizational closure. *Organization*, *11*(6), 793–824.

Cunliffe, A. L. (2002). Social poetics as management inquiry: A dialogical approach. *Journal of Management Inquiry*, *11*(2), 128–146.

Czarniawska, B. (2013). Organizations as obstacles to organizing. In D. Robichaud & F. Cooren (Eds.), *Organization and organizing: Materiality, agency, and discourse* (pp. 3–22). New York: Routledge.

Deetz, S., & Simpson, J. (2004). Critical organizational dialogue: Open formation and the demand of "otherness." In R. Anderson, L. A. Baxter, & K. N. Cissna (Eds.), *Dialogue: Theorizing difference in communication studies* (pp. 141–158). Thousand Oaks, CA: Sage.

Dervin, B., Foreman-Wernet, L., & Lauterbach, E. (2003). *Sense-making methodology reader: Selected writings of Brenda Dervin*. Creskill, NJ: Hampton.

Eisenberg, E. M., Goodall, H. L. Jr, & Trethewey, A. (2013). *Organizational communication: Balancing creativity and constraint* (7th ed.). Boston, MA: Bedford St. Martin.

"Emails Show Casual Response by GEMA Director Over Winter Storm." (2014, February 3). *WSB-TV Atlanta report*. Retrieved from http://www.wsbtv.com/news/news/local/emails-show-casual-response-gema-director-over-win/ndBwf/

Fairhurst, G. T., & Cooren, F. (2009). Leadership as the hybrid production of presence(s). *Leadership*, *5*(4), 469–490.

Governor's Severe Winter Weather Warning and Preparedness Task Force Report. (2014). Retrieved from gov.georgia.gov/sites/gov.georgia.gov/files/related_files/press_release/Governor%27s%20Severe%20Winter%20Weather%20Warning%20and%20Preparedness%20Task%20Force%20Report_4.3.2014.pdf

Holman, D. J., & Thorpe, R. (2003). *Management and language: The manager as practical author*. London: Sage.

Lorenzoni, I., & Pidgeon, N. F. (2006). Public views on climate change: European and USA perspectives. *Climatic Change*, *77*(1–2), 73–95.

Luhmann N. (1982). The world society as a social system. *International Journal of General Systems*, *8*(3), 131–138.

MacMath, J. (2014, December 22). What is a polar vortex? *AccuWeather*. Retrieved from http://www.accuweather.com/en/weather-news/what-is-a-polar-vortex/21793077

McCaffrey, S. (2014). Second ice storm gives Gov. Deal a 'do-over.' *The Atlanta Journal-Constitution*. Retrieved from http://www.myajc.com/s/news/local/snowstorm/

McNamee, S., & Shotter, J. (2004). Dialogue, creativity, and change. In R. Anderson, L. A. Baxter, & K. N. Cissna (Eds.), *Dialogue: Theorizing difference in communication studies* (pp. 91–104). Thousand Oaks, CA: Sage.

McPhee, R. D., & Zaug, P. (2000). The communicative constitution of organizations: A framework for explanation. *Electronic Journal of Communication*, *10*(1–2), 1–16.

Mumby, D. K. (2012). *Organizational communication: A critical approach*. Thousand Oaks, CA: Sage.

Murphy, R. (2004). Disaster or sustainability: The dance of human agents with nature's actants. *Canadian Review of Sociology & Anthropology*, *41*(3), 249–266.

Payne, E. (2014, February 10). Ready, says Georgia. *Cnn.com*. Retrieved from http://www.cnn.com/2014/02/10/us/winter-weather/

Plotkin, J. (2014, January 28). *Winter storm Leon roars through the south*. Retrieved from http://www.decodedscience.com/winter-storm-leon-roars-south/42226

Putnam, L. L., & Nicotera, A. M. (2009). Communicative constitution of organization is a question: Critical issues for addressing it. *Management Communication Quarterly*, *24*(1), 158–165. doi:10.1177/0893318909351581

Schön, D. A. (1983). *The reflective practitioner: How professionals think in action*. New York: Basic Books.

Scott, M. B., & Lyman, S. M. (1968). Accounts. *American Sociological Review*, *33*(1), 46–62.

Searle, J. (1969). *Speech acts: An essay in the philosophy of language*. Cambridge: Cambridge University.

Severe Winter Weather Warning and Preparedness Task Force. (2014, April 3). Final report. Retrieved from http://gov.georgia.gov/sites/gov.georgia.gov/files/related_files/press_release/Governor%27s%20Severe%20Winter%20Weather%20Warning%20and%20Preparedness%20Task%20Force%20Report_4.3.2014.pdf

Shotter, J. (1993). *Conversational realities: Constructing life through language*. London: Sage.

Shotter, J. (2004). Expressing and legitimating "actionable knowledge" from within "the moment of acting." *Concepts and Transformation*, 9(2), 205–229.

Shotter, J. (2006). Understanding process from within: An argument for "withness"-thinking. *Organization Studies*, *27*(4), 585–604.

Taylor, J. R., & Van Every, E. J. (2000). *The emergent organization: Communication as its site and surface*. Mahwah, NJ: Lawrence Erlbaum Associates.

Weick, K. (1979). *The social psychology of organizing* (2nd ed.). New York, NY: Random House.

Weick, K. E. (1995). *Sensemaking in organizations*. Thousand Oaks, CA: Sage.

Weick, K. E. (2001). *Making sense of the organization*. Malden, MA: Blackwell.

Wilson, J., & Smith, M. (2014, February 1). Two inches of chaos: What Atlanta's leaders were doing while the snow fell. *CNN Report*. Retrieved from http://www.cnn.com/interactive/2014/02/us/georgia-snow-timeline/index.html

Winograd, D. (2014, January 30). Georgia governor apologizes for Atlanta snow gridlock: Admits state was not prepared for storm. *Time* (online). Retrieved from http://nation.time.com/2014/01/30/georgia-governor-apologizes-for-atlanta-snow-gridlock/

· 5 ·

CONSTITUTING AGENCY

Taking Action (or Not) During the Hurricane Katrina Disaster

Agency is a significant topic in the CCO approach given the relationship between agency and action. It is also a problematic concept given that there has been a great deal of disagreement across the main CCO approaches on what is agency, and who *or what* can be an agent. As a starting definition, agency can be described as the capacity to make a difference (Cooren, 2010). The Montreal School considers that humans *and* "nonhumans" can be agents or have agency. In other words, not just people, but also objects (e.g., copying machines, post-it notes, speed bumps, etc.) have the capacity to make a difference and therefore, can be agents in affecting organizational action.

This chapter explores how agency is constituted through communication. This is an important issue for organizational communication scholars and practitioners generally, but especially during a crisis in that agency is linked to action: there are several situations where something should or must be done. To illustrate how agency is communicatively constituted and the importance of this process, an extended case study analysis of teleconferences that took place during the Hurricane Katrina disaster among Louisiana emergency managers will be analyzed. During this situation, in the moments of interacting, there was a great deal of uncertainty regarding the hurricane and its impact and how to mobilize for responding to the crisis situation. This situation is also a rich one for examining the role of humans and nonhumans in responding to a crisis given the dependency of human agents upon nonhuman entities (i.e., generators, roads, shelters). This chapter addresses the issues of how is organizational agency enacted during a crisis, and what is the role of humans and nonhumans in constructing a crisis.

Agency is a socially and culturally constructed construct (Ahearn, 2001). An implication of this statement is that agency is relationally-mediated: who or what has agency depends on relationships. As a result, how humans and nonhumans play a role in the communicative constitution of crises depends on their relationships; these are not necessarily determined beforehand, but rather, are worked out in moments of interaction. In other words, agency is not an inherent quality of humans or nonhumans; rather, agency is communicatively constituted in interaction where participants may disagree about, apply, promote or demote *who or what* counts as an agent.

The featured analysis of this chapter focuses on key accounts during teleconferences among emergency managers as Hurricane Katrina approached Louisiana. The type of discourse that is analyzed to discuss agency during these teleconference meetings are "accounts" which are linguistic devices offered when someone does something wrong. The "wrongdoing" could range from an everyday occurrence such as showing up late for lunch to something more major and with broader impact such as a major malfunction with your company product that resulted in lost lives. When these "wrongs" happen, one way that people cope with them is through communication. Broadly speaking, they may offer an account for the action. More specifically, they could apologize, make an excuse, blame someone else, or deny that something wrong occurred.

Accounts implicate agency in that they can make action accountable by repairing social breaches, and in doing so, construct notions of responsibility or who (or what) was responsible for what action (Buttny, 1993; Buttny & Morris, 2001; Scott & Lyman, 1968). Accounts are a rich discursive resource for examining how agency is constructed during crisis response meetings. In the remaining sections of this chapter, key views of agency are described, a model of agency as discursively mediated is explained, and an analysis of Hurricane Katrina teleconferences is provided. The teleconferences can be considered an instance of failed agency in that inadequate actions were taken to prepare for and respond to Hurricane Katrina. As such, the teleconferences provide a rich set of data to examine how agency is enacted in moments of interaction rather than as something that is pregiven ahead of time.

Theorizing Agency

Agency has been an important yet problematic issue in scholarly research generally and organizational communication research specifically. Two dominant approaches in this area include a structuration approach, derived from Giddens' (1984) work in sociology, and Latour's (2005) actor-network theory based on the sociology of science; these approaches correspond with what has been labeled internalist and externalist views of agency (Robichaud, 2005). In the remainder of this section, an overview of these two dominant views of agency is provided and current controversies in theorizing agency are explained. Then the idea of agency as a social construction that is historically, relationally, and culturally situated is addressed.

Internalist and Externalist Views of Agency

There are several philosophical views of agency such as agency as free will, as resistance, and as absence (Ahearn, 2001). One contrast that has been described within the Communication discipline has been that between the intentionalist and externalist views of agency (see Robichaud, 2005). The tension between these views highlights key controversies and challenges for scholars in theorizing agency and so explaining these two perspectives will be the focus within this section.

The *internalist view* is best represented by Giddens' (1984, 2013) structuration approach in which agency and structure exist in an interconnected relationship described as a *duality of structure*. Structuration theory is a key, root theory of the Four Flows CCO approach, and so the view of agency advanced through structuration theory is aligned with the Four Flows approach.

In the idea of the duality of structure, agents may be shaped by and constrained by structure in the form of rules. But agents in turn have the capacity to act back upon and alter rules and structure. Thus, agents are intentional and reflexive (McPhee, 2004). In the internalist view, agents are assumed to be able to think consciously and intentionally about their actions and choices.

When actions are repeated, also known as iteration, they become institutionalized and part of a structure that is followed in future actions. The internalist view has been associated with Berger and Luckmann's (1966) early social constructionist work in that the internalist view relies upon human habits and routines for the maintenance of structure (Robichaud, 2005). However, neither the internalist nor externalist views map neatly onto social constructionist views as there are connections across both views to social constructionism, especially given the diversity of perspectives within social constructionist theorizing.

Another way of thinking about the internalist view is that it assumes that agency is located *within* the individual: we as individuals think about and make choices. In this process, we exercise our agency.

There are a number of critiques of the internalist view. Robichaud (2005) points out that the internalist view of agency conflates the agent with a human actor. There are two parts of this counter-argument. First, agent and actor are not equivalent in that agent relates to action whereas actor relates to the nature of one's being as a human. In other words, just because one is a human does not automatically endow one with a capacity to act. There are times when people are not able to take action because other people or other things may be stopping them. Second, the internalist view assumes that only humans can be agents thereby failing to account for the role of material, nonhuman objects in constituting action. Actions may occur without humans or may be out of the control of humans. For example, consider how your entry into or out of a parking structure is controlled by ramps, signs, and automated barriers. The externalist view addresses these critiques.

The *externalist* view is characterized by the notion of agency as "composition" or as hybrid (Callon & Law, 1995). This perspective is best represented by Latour's (2005) Actor-Network Theory (ANT) which is based on his work in the sociology of science. I highlight this relationship as it emphasizes how ANT needed to deal with the materiality of objects and of the world in participating in science, scientific discovery and the development of scientific knowledge. Also, ANT is one of the key influencing theories in the development of the Montreal School of CCO thought.

In advancing a hybrid notion of agency, Latour, along with theorists in the Montreal school (see Cooren, Taylor, & Van Every, 2005; Robichaud & Cooren, 2013), argued for a view of agency that includes nonhuman agents or actants. The terminology of actant in lieu of agent functions to bypass the conflation of actor with agent in that an actant may consist of collectives (i.e., "The government declared …") or even nonhumans (i.e., "The car swerved"). Scholars subscribing to the externalist view focus on the primary defining feature of agency as the capacity to "make a difference", and therefore nonhumans such as texts and other material objects may be considered actants (see Pickering, 1993). Because objects may act, this view has also sometimes been called a posthuman view.

This perspective is not without its controversy and criticisms. Intentionality and reflection are missing from a posthuman view of agency in that nonhumans do not have intentions or the capacity to reflect upon their actions (McPhee, 2004). In addition, nonhumans do not have the ability of rationalization or to make their actions accountable (McPhee, 2004).

The internalist and externalist views each have compelling arguments regarding agency as well as challenges in conceptualizing agency. Some of the problematic issues include how to conceptualize and explain collective agency, how to account for agency from afar (i.e., local versus translocal), how to account for the roles of texts, nonhumans, and the material in the constitution of organizing. In the remainder of this chapter, I propose a communication social construction perspective of agency. I seek to address the aforementioned topics on agency by foregrounding the role of discourse or situated-interactions in the communicative constitution of agency.

Agency From a Communication Social Construction Perspective

Our society's current view of agency is a modern one, arising from the Enlightenment period and more recently, from changes in political and economic systems (Meyer & Jepperson, 2000). As Meyer and Jepperson (2000) explained:

> [agency] is also a construction in a more specific sense: its elaboration is tied to the liberal model of sociopolitical organization that has dominated the post-World War II epoch, with its emphasis on individual rather than diffuse corporate or state authority, on democratic forms and the market economy, and so on. (pp. 108–109)

Within a social construction perspective, "things" may be constructed as having agency. For example, through various legal and linguistic proceedings, the "corporation" is treated as an actant. However, as Meyer and Jepperson (2000) argued, "for an entity with recognized interests to be seen as a legitimate actor requires another step: the cultural construction of the capacity and authority to act for itself" (p. 105). *How* the construction of an entity as agent occurs is a communication process wherein constructions of agency are relational, discursive, and socially accountable.

Robichaud (2005) articulated a *relational view of agency*. For this, he utilized American philosopher Charles Sanders Peirce's triangle of meaning and the example of gift-giving. In order for a gift to be given, a number of conditions must be met: there has to be a giver, there has to be a gift, there has to be a gift recipient, and these must be oriented to each other in the gift-giving process. Robichaud's treatment highlights how agency is not a solo endeavor but relies on the linguistic and social cooperation of others and in this respect, agency is relational.

The interpretation of gift-giving is also based on discursive, social, and historical resources. For example, using again the gift scenario, what a gift means is based on the following assumptions: the desirability of the gift, a kind intention associated with a gift, transfer of ownership upon the giving of the gift, etc. There are a number of examples (mostly drawn from the interactions of children) that can illustrate the failure of gift-giving at any of these stages. Most children would not view receiving coal from "Santa" as a gift. Many children (or adults for that matter) would not view receiving a spider in a box as a gift. And, when gifts are given, they are not typically "borrowed" back.

Czarniawska (2013) addressed the issue of agency through *action-nets* to focus on "outcomes rather than inputs of organizing" (p. 13). Intentionally replacing the vocabulary of "network" with "net", Czarniawska explained that

> Network assumes the existence of actors, who forge connections. Action net reverts this assumption, suggesting that connections between and among actions, when stabilized, are used to construct the identities of actors. One becomes a publisher when one starts to publish books or journals, which means that connections have already been made with such actions as writing and printing. (p. 14)

In other words, it is the connections or relationships that create the actors and their capacity as agents. Czarniawska argued that in replacing actor-networks with action-nets, the key factor in constructing agency is based on time or process in that it is ultimately action or activity rather than a passive connection that constitutes agency. Similarly, Pickering (1993) argued for a view of agency that described agency as temporally emergent. Like Robichaud (2005), his view is a semiotic one that also noted the shifting character of agency: agency is *not* the inherent property of someone or something but emerges in practice (Pickering, 1993).

Activity or "temporal emergence" is constituted by communication practices. As Ahearn (2001) explained, language and agency are interconnected in that constructions of agency are built into the grammatical structuring of language (also see Duranti, 1990). Ahearn described anthropologist Keith Basso's (1996) analysis of the importance of place among the Western Apache of the U.S. as identified through stories about place. In this example, agency is shaped through tales that are told of the landscape (see Carbaugh, 1999).

Ahearn's point regarding agency and space among Apache is that their conceptualization of agency is shaped by cultural assumptions and cultural ways of speaking about agency. Basso highlighted how, for the Apache, nonhumans such as places and landscapes may enact agency. Paraphrasing from others, Ahearn described this as agency extending "beyond the skin." In some cultures, one is an agent or may enact agency through such means as witchcraft (see Scott & Lyman, 1968). Or, one may lose his/her agency because of an action and chain of events that would not make sense in other cultures. For example, within the Old Testament, there is the story of Job to whom many bad things happen as part of a challenge of Satan to God. A consequence of this challenge is that Job has lost his agency.

What constitutes agency is culturally up for grabs, and therefore, in some cultures it may make sense to talk about the agency of places, objects, and/or invisible entities whereas in other cultures it would not.

Materiality and Agency

The preceding examples of the construction of places as agents indirectly brings in the issue of materiality and agency. Materiality is especially important for this project given the undeniability of physical characteristics of natural disasters such as hurricanes with their powerful winds, torrential rains, and consequences such as the loss of lives, homes, and livelihoods. One way that the argument about materiality has been handled is to shift to focusing on how meaning is created through communication (Edwards, Ashmore, & Potter, 1995). The following section advances a notion of agency that includes discursive and material factors and focuses on how these socially construct agency in a collaborative way.

While maintaining the agency of *both* humans and nonhumans, Murphy (2004) used the analogy of dance: "Human agents dance with the moves of nature's actants to form hybrid constructions, with both influencing the other and both having some autonomy" (p. 254). In some instances, humans may take the lead in the dance whereas in other instances, nature's actants may take the lead. Murphy distinguished these different types of agencies by noting that humans are purposive agents whereas non-humans are actants who may exert agency albeit unintentionally. Murphy utilized the case study of Canada's 1998 ice storms which was the most financially costly natural disaster in Canada's history by overwhelming Quebec's electrical system resulting in the loss of electricity with associated consequences such as loss of light and heating for people, failure of cooling systems to preserve food, and loss of needed energy for dairy farm maintenance. Murphy argued that this disaster was a socially constructed one in that ultimately, the disaster was caused through Quebec's reliance on specific technological systems. To highlight and illustrate this point, Murphy contrasted how the Amish of New York state handled the same ice storm: they were basically unaffected. As Murphy summarized:

> The comparison of Quebec society with the Amish demonstrates that disaster resulted not from freezing rain per se, but rather from the vulnerability of the infrastructure that modern society had constructed and upon which it had become dependent. (p. 257)

This is not intended to blame let alone cast an unsympathetic eye toward Quebec and its ice storm disaster; rather, this is intended to highlight the complexity of describing the cause of a disaster: the constitution of the ice storm as disaster occurred through a social construction process. Humans and nonhumans (the ice storm, technological systems) exerted agency in varying ways but in the manner of a dance: the disaster would not have occurred without the ice storm *and* without human reliance upon particular technologies.

Similar to the idea of "dance," Pickering (1993) utilized the term "mangle":

> "Mangle" can also be used as a verb: I want to say, for example, that the contours of material agency are mangled in practice, meaning emergently transformed and delineated in the dialectic of resistance and accommodation. (p. 567)

Pickering also argued that material agency is "temporally emergent in relation to practice so conceived" (p. 565) and that "material and human agencies are mutually and emergently productive of one another" (p. 567). Because of the close relationship between humans and tools in the enactment of agency or in holding agentic capacity, Wertsch, Tulviste, and Hagstrom (1996) used the phrase "mediated agency" to capture the hybrid dynamics of agency: "instead of the individual in isolation, the agent is viewed as being an irreducible aggregate of individual (or individuals in intermental functioning) *together with mediational means*" (p. 341). Similarly, humans and nonhumans can be viewed as functioning as hybrid agents in working together in effecting action.

Summary

In advancing a communication social construction conceptualization of agency as discursively-mediated, I propose the following characteristics regarding agency:

- Agency is co-constructed in and through language-in-interaction and in this respect, is relational.
- Agency is situationally-based where situation may include history and cultural assumptions.
- Because of the preceding characteristics, humans and nonhumans are candidates for exercising agentic powers.

Case Study: Hurricane Katrina

This analysis[1] focuses on teleconferences among parish, state, regional and federal emergency managers that occurred during the Hurricane Katrina disaster. In a way, this particular case illustrates the failure of agency and in doing so, allows for discussion of how communication operates in the constitution of agency and the conditions and ways in which it does so. There were ten teleconferences that occurred among Louisiana local (i.e., parish leaders), state (i.e., representatives from health and human services and transportation units), and federal officials (i.e., National Weather Service and FEMA representatives) as Hurricane Katrina approached. The purposes of the teleconferences were partly informative in providing updates and partly coordinating in reporting emergency needs. However, as with what generally occurred during the Hurricane Katrina disaster, coordinating efforts within these teleconferences were not successful. The teleconferences were recorded by Jefferson Parish Emergency Manager Walter Maestri and were later made available to the public.

Ahearn (2001) argued for looking at the interrelationship of language and agency, pointing out that agency is embedded in language structure, while also noting that agency is socially constructed. Accounts are one such language practice that can provide insight into the construction of agency. The analytic unit of analysis consists of accounts offered during the teleconferences. Accounts are linguistic devices that are offered during problematic situations (Scott & Lyman, 1968). There are a range of ways to conceptualize accounts that vary from the direct offering of an excuse or justification when accused of a wrongdoing (see Scott & Lyman, 1968) to making one's behavior "accountable" or understandable along the lines of an ethnomethodological approach (see Garfinkel, 1967; also see Buttny, 1993; Buttny & Morris, 2001).

During the teleconferences, there are three episodes of particular interest for this chapter. The first episode is a situation in which after a teleconference (the second teleconference for the meeting group) was officially declared over, participants continued on the line to talk. This episode features the negotiation of human and textual agency. The next two episodes feature FEMA, with one episode highlighting how physical presence does not equate with agency and the latter highlighting the role of competing activity nets in the management of the emergency response.

Negotiating Human and Textual Agency

This first example is taken from talk that occurred after the second teleconference was officially over. For this teleconference, the moderator had concluded the call, but some participants attempted unsuccessfully to gain the moderator's attention to pose additional questions. The result was that the moderator and many others had exited the call, but some continued on the line to pose questions and answer questions for each other. Because this talk was not regulated by the same formal guidelines that were applied for the official teleconference, speakers were able to pose follow-up questions and speak more informally.

The following excerpt illustrates a clash in authority between human agents and textual actants. The exchange begins with a participant asking about when the "phase one" evacuation was going to be declared. From there, she made an inquiry about the evacuation process and where a particular shelter was going to be located. Italics have been added to emphasize certain words and phrases for analytic purposes.

Excerpt 1 (8/27/2005 7:30 a.m.):

1 Woman: *He said* (inaudible) at some kind of truck stop.
2 Woman 2: T and A stop.
3 Man: T and A.
4 Woman: *This says* (inaudible) Welcome Center ...
5 Man: It's under construction, that's *what the girl said*, that's the reason why they ...
6 Woman: [BUT WHY ARE WE (inaudible]
7 Woman2: [MY PROBLEM IS, wait a minute. MY PROBLEM IS they *tell* me to
8 *read* this DAMN *thing and it's not the same thing*.
9 Man: No, you heard what *the guy* said.
10 Woman2: Well, we're gonna *call the state* right now.
11 Man: You heard, [he said it], it's under construction, can't you tell it's a more
12 specific point, and that's why *they said* TA Truck Stop.

There are multiple human agents and nonhuman actants that are being used or "mobilized" in a type of duel for authority in attempting to figure out what was communicated about the evacuation procedures during the teleconference. In line 1, the "woman" meeting participant reiterated information that she thought she heard during the meeting about the location of a shelter. That information came from an unspecified "he." In this instance, the "he" would be considered an actant in declaring that a truck stop would be used as a shelter. In line 4 however, another interlocutor pointed out that a different site (a Welcome Center) was identified in written documentation as the shelter location ("*this says*"). In line 5, the "man" offered "reasons" and the words of a "girl" to counter the authority of the written document in identifying the shelter area. In this chain of legitimation, the "girl" spoke on behalf of the "reason" which took priority over the authority of the "this" (written documentation).

In lines 7 and 8, the "woman 2" speaker explained a key problem. On the one hand, she was told to read a document as a source of information or authority regarding what to do during the hurricane emergency situation ("*they tell* me to *read* this *DAMN thing*"). Therefore, certain human agents ("they") have given authority to a specific organizational text referred to as the "DAMN thing." The text of course is an expression of other human agents since someone wrote the evacuation plan. However, it is the written document rather than the author or authors of the document that is referenced by the speaker as the authoritative voice. In this respect, this exchange also illustrates Cooren's Ventriloqual approach as various organizational entities are made to speak on behalf of others.

However, the document is not accurate for the current conditions that the emergency managers must deal with ("it's not the same thing"). Borrowing from terminology from Actor Network Theory, one way to view this circumstance is that the document (emergency plan) has "betrayed" them in its lack of accuracy. Alternatively, it can also be said that the humans involved in emergency planning betrayed their documentation by not keeping it updated.

In describing the Montreal school approach, Putnam (2013) explained, "texts reflect what was said and provide a record of past conversations; thus, conversations move through time and space in the form of texts" (p. 26). In this instance, a text (emergency plan) was created based on conversations that were in anticipation of a projected future. The texts, in other words, were created in anticipation of guiding and participating in future conversations that would be occurring during a hurricane emergency situation. However, the authority of the text was questioned because of its inaccuracy in relating to the current, unanticipated circumstances of Hurricane Katrina, which in this excerpt is the lack of readiness of the initially designated shelter.

Ultimately, the more immediate authorities as enacted in the oral discourse took priority. What is also of note here is a conflict of how to follow procedures. The written guidelines were presumably developed as a plan to follow or source of authority in case of an emergency. However, they were undermined by a mismatch in what is specified in writing versus the materiality of the situation at hand. In this case, knowledge that is local, in both the spatial and temporal sense, was more reliable.

What is significant about this excerpt is not so much what the final agent or authority is, but the fact that agency was discursively worked out in the organizing talk of meeting participants. To just focus on what was the dominant agent glosses over the issue that it was not apparent to all participants who or what the key authority was. Agents/actants are not givens, but worked out in discourse through the accounts presented by the speakers. It is this activity that is the substance of the communicative constitution of agency. Presence of supposed "agents" are not enough. Rather, agents may be activated or deactivated based on conversations.

The Non Equivalency of Agency and Human Actors

The Federal Emergency Management Agency's role in the Hurricane Katrina disaster was controversial due to its many inadequacies and failures in responding to the emergency situation (see Cooper & Block, 2006). It can be said then, that FEMA illustrated a failure in agency in both the academic and governmental sense. For the first nine of ten recorded teleconferences, at least one FEMA representative was present and had an opportunity to provide a report during the teleconference. For four of the teleconferences, a report was provided by a FEMA official. In five of the teleconferences, the FEMA official indicated that he had "nothing to report," however, in three of these instances, the moderator (JS) immediately followed-up on the FEMA non-report by providing an alternative account of FEMA activities. The following example illustrates this dynamic:

Excerpt 2 (Aug. 27; 10:30 a.m.):

JS: Mississippi EOC, have you joined us, and if you have, do you have anything
that ah, any comments?. … Mississippi EOC, not online. *FEMA, Region 6*
any information?. … FEMA liaison? Any information at this point?
FEMA liaison: None right now.
JS: Again, *it's been reported earlier,* FEMA *(inaudible) 8 team is on the way*

and we also have a FEMA (urg?) 10 team that's on the way. So we, we have
ah, FEMA assistance that's ah, rolling in, and we have a liaison here at this
particular point in time. We will now go to our parish roll call so have you, ah,
to give you an opportunity to ask questions.

In the above excerpt, when given an opportunity to speak, the FEMA liaison stated that he had no comments "right now." Even though a FEMA representative was present on the line and reported that he had no information to provide, in the turn immediately following, JS acted as a type of representative for FEMA by providing a report in specifying what FEMA teams were on their way. The FEMA liaison's account was minimal; his comment, "none right now," implied that nothing was going on, but something worthy of describing may occur later. However, JS' follow-up comment can be viewed as a negative assessment of the account offered by the FEMA liaison in that JS himself provided an account of FEMA activities by supplying details not stated by the present FEMA liaison.

Cooren's (2010) Ventriloqual approach to agency can be applied here to describe what is going on. In this example, we could say that the FEMA representative's account was inadequate due to its minimal information and therefore, JS stepped in with an account and acted as a "figure" allowing FEMA to speak through him to provide details. The FEMA representative did not successfully bring FEMA into the presence of the meeting (see Benoit-Barné & Cooren, 2009). The comparison of these two FEMA reports, one from the official FEMA representative and the other from JS, illustrates how being present is not sufficient for the enactment of agency.

One counter-argument is that perhaps the FEMA liaison was ignorant himself of FEMA activities and that is why he had no report. Or perhaps, he was unaware of what type of details and activities would be appropriate to report in that setting. What's important here in this comparison of the two FEMA-related reports is that the presence of an agent or of one associated with an agency is not sufficient for the enactment of agency. Agency is not an inherent aspect of an individual or organization, but rather a capacity that can be "built up" in specific conversational activities.

Competing Action Nets

The preceding example focused on the presence of FEMA during the teleconference when there was also a person present who is representing FEMA. The next example is taken from a circumstance in which a FEMA representative is not physically present, and that actually is part of the crux of the problem being discussed. The following excerpt is from a teleconference that occurred approximately a week and a half after the hurricane. At this point, the focus of the teleconference is on relief efforts, surveying damage and getting help to those who did not evacuate. The human disaster is becoming apparent as the anticipated emergency relief efforts are falling apart with confusion and delays.

In the excerpt below, the Jefferson parish representative reflected on the problems in trying to manage response and emergency relief efforts:

Excerpt 3 (Sept. 9):

Jefferson: Yeah, ah Jeff, this is Walter. Um, *we have that same problem with the right hand not knowing what the left hand is doing. Ah, I don't know how we share* with these people what our plans are and how the parish plans meld with the state plans and all resources ought to be cleared through the EOC's because ya know, *people are bypassing the system* and, and different, different resources and assets are being sent to places that ah, you know, we know nothing about. *And that's, that's a real problem right now, trying to manage this situation.*

JS: Yeah, yeah W. I can, I concur a hundred percent ...

The preceding remarks highlight some of the challenges in coordination that were occurring during the Hurricane Katrina response and relief efforts. The Jefferson representative located a key problem in coordination ("the right hand not knowing what the left hand is doing"). Behind this problem or a cause of this problem is that "people are bypassing the system." In other words, the leaders of the parish emergency response groups are being excluded from hearing information about and decisions on resource requests and allocations. JS highlighted another aspect of the failure of coordination in that the federal coordinator officer was not working sufficiently with other groups.

The account of the problem provided by the Jefferson parish representative highlights competing action nets (Czarniawska, 2013) vis-a-vis the dilemma of "melding" parish and state plans. There is supposedly a system of working through the EOC, but as the Jefferson representative highlighted, "people are bypassing the system." In Czarniawska's conceptualization of action nets and Robichaud's (2005) discussion of a relational approach to agency, both recognized that the enactment of agency is contingent upon reciprocation or "activity" as Czarniawska would explain.

The next example begins with the Jefferson representative describing a problem with obtaining food and water:

Excerpt 4 (September 9):

Jefferson: Um … we've got some food issues as well. Both, both quantities and distribution. Um, you know, in the beginning food was pouring, food and water were pouring in, now it's trickling and you know, we have no, no supermarkets opened, no place for people to buy food, water, or whatever, and we need to get that, you know, we need to get more food into the, into the parish as quickly as possible so I can get it to the distribution site.

JS: Now again, I assume you are talking about the emergency food, the water, the MREs and the ice?

Jefferson. Correct, that's exactly right.

JS: And is that, ah been requested and turned down or are you just requesting it now?

Jefferson: No, we, we, we requested it, have we requested it, been requesting it, and nothing's coming.

The above excerpt can be described through a variation of Robichaud's gift giving example. In Robichaud's example, a gift is given and a gift is received and in this exchange, an action successfully occurs. The gift giver can be said to have agency, mainly through the cooperation of the gift receiver in receiving the request. In the above example, there are requesters, but "nothing's coming." This is akin to giving a gift that is denied, but instead, a request is being made and ignored. With the situation of gift giving, for this to occur, there are a number of other actions that must occur that include receiving the gift. One can make a request, and indeed, the linguistic nature of requests is that the other has some options or choices in how to respond. In the example from the teleconference meeting, the responses were sometimes "it's coming," but in actuality, did not arrive, at least not in a timely manner. This scenario highlights how the enactment of agency is dependent upon other relationships, and without cooperation, agency as the capacity to act, falters.

The next series of excerpts are taken from the same teleconference call. While there are several topics discussed, the following comments focus on coordination with the FEMA office:

Excerpt 5 (September 9):

JS: Okay Walter, ah, did you hear about the daily FEMA conference call, it's ah, that the liaison's have been having at the joint field office?

Jefferson: Right, I filled out his form with him yesterday and uh, and uh, he didn't tell me about the conference call but I'm gonna find out about it, and I'm gonna try and join that call.

JS: Yeah, the, the minute I heard about it uh, we realized how every, how busy

everybody is, and instead of trying to have a separate state call and so forth I think
doing the one call will save everybody's time, because I'm sure now that
communications are up, if you've got a critical issue you know to please call me.

While there are several problems associated with the hurricane relief efforts and coordinating that work, one of the problems highlighted in the above excerpt is that FEMA has basically set-up an alternative, potentially competing, "action net" (Czarniawska, 2013). While the coordinating calls between the parish leaders were occurring, FEMA had set up an alternative set of conference calls to try to coordinate. As the speakers in the preceding excerpt highlighted however, this created inefficiencies in communication and, at the same time, cut out individuals in crucial roles such as the parish emergency response managers.

Summary

The analysis focused on specific moments where organizing broke down, or, moments where there was a breach in social order and someone was called upon to provide an account. In doing so, this chapter was intended to illustrate how agency is constituted through communication.

The examples discussed from the teleconferences showed different ways that participants discursively negotiated agency during the crisis situation. The first example illustrated how agency may be exercised by people or texts as the participants attempted to reconstruct the instructions provided on the evacuation procedures. The different agencies at play consisted of the teleconference mediator, state traffic control officials, and the evacuation plan. While the directions from the state officials eventually took priority, what is of note in this example was that textual authority was taken seriously and that there needed to be reasons provided for deviating from the plan.

The second example of the FEMA report during the teleconference demonstrated how agency should not be treated as the same as being a human; while the FEMA representative provided an account regarding FEMA actions, someone else not from FEMA stepped in to provide a more thorough accounting of FEMA actions. The third example highlights the role of action-nets and how competing action-nets can stifle action: locally-articulated needs and requests were being ignored as the FEMA-managed net was being set-up.

Texts have the potential to act as agents or actants, but they, in and of themselves, are not actants. In order to operate as an actant, texts are dependent upon other social practices. We may treat them as actants, but this may just be a temporary matter as, at any given point someone may point out that the text is just a piece of paper or is not suitable. Granted, written texts do have a sense of permanence, but the permanence of what a text means is dependent on the cooperation of humans. Take as a case in point the role of the emergency plan in responding to Katrina. This was not kept up to date with the changing road construction conditions in Louisiana and thus was not an effective agent in the hurricane relief efforts.

Conclusion

Through this chapter, I wished to address current debates in CCO on the nature of agency. A prime tension in this area has related to the role of nonhumans as agents. I elaborated on a discursively-mediated approach to agency to account for the variety of "things" that may play agentic roles while also addressing the different types of relationships that humans and nonhumans may have. Human agents and nonhuman actants do not play absolute roles in how they can make a difference or influence action, but rather, their activities are discursively mediated. The analysis of Hurricane Katrina teleconferences illustrated the different dynamics of agency as discursively mediated ranging from how the agency of a text (evacuation plan) can be discursively-negotiated, to how a collective agency (FEMA) may exert agency through various spokespersons, to the role of competing action-nets in developing a crisis response. These cases illustrate the relational nature of agency as activity rather than as an inherent property of anyone or thing.

In focusing on Hurricane Katrina, I addressed a situation that emphasizes the importance of agency and of Pearce's (1994) call for "recovering agency." This case demonstrates how human agents and nonhuman actants interact in affecting action. These interactions are akin to the "dance" described by Murphy (2004), however, what this study highlights is how these interactions are discursively-mediated and relationally-constructed (also see Cooren, 2006).

The Montreal School approach emphasizes that agency is constituted in the capacity to make a difference. This capacity can be applied to humans and nonhumans. With the example of Hurricane Katrina, there are several ways that the idea of agency can be applied. In one way, the hurricane itself can be considered an agent in that it "made a difference" through its strong winds and torrential rains. However, it is also important to note that the impact of Hurricane Katrina was not due just to the hurricane itself, but also to the specific characteristics of the social and infrastructural systems that it affected and conversely, the subsequent responses from social systems.

Note

1. I wish to thank my research collaborator and friend, Mariaelena Bartesaghi for inviting me to work with her on a research project to study the Hurricane Katrina teleconferences. Mariaelena obtained the transcripts that were then analyzed for this chapter.

References

Ahearn, L. M. (2001). Language and agency. *Annual Review of Anthropology*, *30*, 109–137.

Basso, K. H. (1996). *Wisdom sits in places: Landscape and language among the Western Apache*. Albuquerque, NM: UNM Press.

Benoit-Barné, C., & Cooren, F. (2009). The accomplishment of authority through presentification how authority is distributed among and negotiated by organizational members. *Management Communication Quarterly*, *23*(1), 5–31. https://doi.org/10.1177/0893318909335414

Berger, P. L., & Luckmann, T. (1966). *The social construction of reality*. Garden City, NY: Doubleday.

Buttny, R. (1993). *Social accountability in communication*. London: Sage.

Buttny, R., & Morris, G. H. (2001). Accounts. In W. P. Robinson & H. Giles (Eds.), *The new handbook of language and social psychology* (pp. 285–301). New York, NY: Wiley.

Callon, M., & Law, J. (1995). Agency and the hybrid collectif. *South Atlantic Quarterly*, *94*(2), 481–507.

Carbaugh, D. (1999). "Just listen": "Listening" and landscape among the Blackfeet. *Western Journal of Communication*, *63*(3), 250–270.

Cooper, C., & Block, R. (2006). *Disaster: Hurricane Katrina and the failure of Homeland Security*. New York, NY: Times Book.

Cooren, F. (2006). The organizational world as a plenum of agencies. In F. Cooren, J. R. Taylor, & E. J. Van Every (Eds.), *Communication as organizing: Empirical and theoretical explorations in the dynamic of text and conversation* (pp. 81–100). Mahwah, NJ: Lawrence Erlbaum Associates.

Cooren, F. (2010). *Action and agency in dialogue: Passion, incarnation and ventriloquism*. Amsterdam: John Benjamins.

Cooren, F., Taylor, J. R., & Van Every, E. J. (Eds.). (2005). *Communication as organizing: Practical approaches to research into the dynamic of text and conversation*. Mahwah, NJ: Lawrence Erlbaum Associates.

Czarniawska, B. (2013). Organizations as obstacles to organizing. In D. Robichaud & F. Cooren (Eds.), *Organization and organizing: Materiality, agency, and discourse* (pp. 3–22). New York, NY: Routledge.

Duranti, A. (1990). Politics and grammar: Agency in Samoan political discourse. *American Ethnologist, 17*(4), 646–666. doi:10.1525/ae.1990.17.4.02a00030

Edwards, D., Ashmore, M., & Potter, J. (1995). Death and furniture: The rhetoric, politics and theology of bottom line arguments against relativism. *History of the Human Sciences, 8*, 25–49.

Garfinkel, H. (1967). *Studies in ethnomethodology*. Cambridge: Polity.

Giddens, A. (1984). *The constitution of society: Outline of the theory of structuration*. Cambridge: Polity.

Latour, B. (2005). *Reassembling the social: an introduction to actor-network-theory*. Oxford: Oxford University.

McPhee, R. D. (2004). Text, agency, and organization in the light of structuration theory. *Organization, 11*(3), 355–371.

Meyer, J. W., & Jepperson, R. L. (2000). The "actors" of modern society: The cultural construction of social agency. *Sociological Theory, 18*(1), 100–120.

Murphy, R. (2004). Disaster or sustainability: The dance of human agents with nature's actants. *Canadian Review of Sociology & Anthropology, 41*(3), 249–266.

Pearce, W. B. (1994). Recovering agency. In S. Deetz (Ed.), *Communication yearbook* (pp. 34–41). Thousand Oaks, CA: Sage.

Pickering, A. (1993). The mangle of practice: Agency and emergence in the sociology of science. *American Journal of Sociology*, 99(3), 559–589.

Putnam, L. L. (2013). Dialectics, contradictions, and the question of agency. In D. Robichaud & F. Cooren (Eds.), *Organization and organizing: Materiality, agency, and discourse* (pp. 23–36). New York, NY: Routledge.

Robichaud, D. (2005). Steps toward a relational view of agency. In F. Cooren, J. R. Taylor, & E. J. Van Every (Eds.), *Communication as organizing: Practical approaches to research into the dynamic of text and conversation* (pp. 101–114). Mahwah, NJ: Lawrence Erlbaum Associates.

Robichaud, D., & Cooren, F. (Eds.). (2013). *Organization and organizing: Materiality, agency, and discourse*. New York, NY: Routledge.

Scott, M. B., & Lyman, S. M. (1968). Accounts. *American Sociological Review*, 33, 46–62.

Wertsch, J. V., Tulviste, P., & Hagstrom, F. (1996). A sociocultural approach to agency. In E. A. Forman, N. Minick, & C. A. Stone (Eds.), *Contexts for learning* (pp. 336–356). Oxford: Oxford University.

· 6 ·

CONSTITUTING SOLUTIONS

Prospective Sensemaking

The preceding chapters and analyses have addressed weather-related circumstances such as hurricanes, winter storms, and heat waves, focusing on failures in how organizations handled crises related to those events. This chapter addresses how organizations could address potential crises to prevent them from happening. The purpose of this chapter is to describe and explain *prospective* sensemaking: sensemaking that occurs to make sense of and anticipate the future. Prospective sensemaking can be very challenging in that it calls for making predictions and thinking through contingencies—basically, attempting to see into the future. There are many terms that can describe activities that organizations already engage in to envision the future such as forecasting, projection modeling, or strategic planning.

Crisis management planning is difficult in that by definition, a crisis is unanticipated, however, crises can be predicted at a general level. For example, I live in the upper U.S. midwest, and so I can anticipate that every winter, there will be at least one blizzard that will result in school and business closings. I know that at the start of the winter season, I should move the ice scraper and snowbrush out of the garage and into the car. Thus at the start of winter, while I and others do not know exactly when a blizzard will occur, we know that it would be wise to be prepared for one to happen at least once during the winter. There are general predictions that can be made as well as preparations to deal with the situation so that it does not reach a crisis state.

The purpose of this chapter is to apply the idea of prospective sensemaking to a situation where action was taken ahead of time to prevent a crisis. The case that will be examined is that of the Great Lakes Compact which was passed in 2008 by Congress and signed into law by President George W. Bush in 2008. In the following sections, an overview will be provided of prospective sensemaking, background for the Great Lakes Compact will be presented, followed by an analysis of congressional discourse during discussion of the Great Lakes Compact.

Prospective Sensemaking

While a great deal of sensemaking research has been retrospective in focusing on how social actors make sense of something that occurred in the past, sensemaking may occur prospectively as people attempt to make sense of an anticipated future (Gioia, Corley, & Fabbri, 2002). Prospective sensemaking has been linked with strategic planning and innovation in that these are activities that are oriented toward the future (Gioia, & Chittipeddi, 1991). However, there are numerous other organizational activities where looking toward the future is required. For example, emergency managers must not only react to a crisis or disaster that has happened, but also must be able to anticipate possible crises and take pro-active measures to prevent such crises. Political leaders also must be able to anticipate and prevent crises.

Weick (1995) argued that "sensemaking can be extended beyond the present. As a result, present decisions can be made meaningful in a larger context than they usually are and more of the past and future can be brought to bear to inform them" (p. 29). Weick (1979) addressed sensemaking of the future through the concept of "future perfect thinking." With future perfect thinking, one thinks in terms of a future that will have happened and because that future is imagined to have happened, one can then work backwards from there to make sense of it. The example that Weick uses is of writing a letter to yourself congratulating you on some accomplishment that you have yet to accomplish as a way of developing a plan to achieve that envisioned future. Overall, Weick's focus and application of sensemaking was retrospective or past rather than future oriented.

A Model of Prospective Sensemaking

Stigliani and Ravasi (2012) developed a model to describe the relationship between retrospective and prospective sensemaking, and individual and collective sensemaking. They identified the following phases of prospective sensemaking: noticing and bracketing; articulating; elaborating; and influence. The first phase, noticing and bracketing bears some similarity to Weick's selection phase. To make sense of something or some event, you must first be able to "see" that it is there. In the second phase, articulating, you begin to develop and express preliminary interpretations to explain what is going on (much like the interpretation phase of Weick's model). In the third phase, elaborating, you go beyond interpreting to making connections. In the fourth phase, influence, you present your sensemaking interpretations and elaborations in a way to influence others so that action could be taken. Stigliani and Ravasi explained that in the influence stage is when organizational members may present claims to clients to persuade them. However, this idea can be generalized to any situation where you are trying to convince someone of something that could happen.

Stigliani and Ravasi's model is similar to Weick's model of sensemaking but assumes a forward rather than backwards looking frame. They noted that prospective sensemaking *is based on* retrospective sensemaking. Their model is also notable and relevant for this project in incorporating linguistic aspects of sensemaking with the role of materiality in sensemaking, a move very much aligned with recent discussions of the communicative constitution of organizing (see Putnam & Nicotera, 2008; Schoeneborn et al., 2014).

Prospective sensemaking research has been applied to strategy making, entrepreneurship, and innovation to "investigate the construction of new understandings of an environment and how to relate to it" (Stigliani & Ravasi, p. 1234). Stigliani and Ravasi contrast a prospective approach with retrospective sensemaking research on crises in that the aforementioned research has shown sensemaking to be "a relatively slow-paced process in which the refinement of emerging interpretations results from cycles of sensemaking and sensegiving, as group members attempt to influence other actors' interpretations (Gioia & Chittipeddi, 1991; Hill & Levenhagen, 1995)" (p. 1234).

Rather than contrast prospective sensemaking research on crises with prospective sensemaking research on strategic planning, this chapter connects a prospective sensemaking perspective to a potential crisis situation thus combining elements of strategic planning and crisis prevention. In doing so, this chapter is intended to provide insight on how risk situations can be identified and proactively addressed *before* they become crises. This approach is compatible with prior sensemaking research findings where scholars have identified how early communication activities prior to a crisis may have contributed to the development of the crisis itself (Weick, 1995).

Discourse and Prospective Sensemaking

Stigliani and Ravasi described prospective sensemaking as occurring in part through the language practices of labeling and articulating. Another example of the relationship between language and prospective sensemaking is from a study conducted by Wright (2005) in which he examined the role of scenarios in prospective sensemaking. He found that "socially constructed stories … integrate predetermined events with critical uncertainties in creative ways" (p. 87). Wright's work illustrates the close connection between prospective sensemaking and how people talk.

Sensemaking and discourse, are interconnected. Cooren (2004) provides one example of the relevance of discourse and sensemaking through his analysis of excerpts from a board meeting to describe how "collective minding" is achieved in conversations through talk that occurred during the board meeting. Cooren's project shows how a sensemaking concept (collective mind) is a social, communication-based activity (collective minding).

There are also specific forms of language that have been associated with sensemaking. For example, scholars have examined how narratives or stories have facilitated sensemaking activities (e.g., Marsen, 2014). Accounts have been featured prominently in analyses of sensemaking especially in the retrospective examination of sensemaking activities. Accounts are also the key area of focus for this chapter analysis in examining account vocabularies for advocating a particular course of action (see Castor, 2005; Geist & Chandler, 1984).

The Great Lakes Compact

The upper Great Lakes of North America, consisting of Lakes Michigan, Superior, Erie, Huron, and Ontario, are one of the greatest resources of freshwater in the world, providing water to tens of millions of regional residents. These lakes are shared within the jurisdictions of Minnesota, Wisconsin, Illinois, Indiana, Michigan, Ohio, Pennsylvania, and New York, in the U.S., and by the province of Ontario in Canada. The Great Lakes are connected with the St. Lawrence River water way, also making Quebec, Canada, a major stakeholder in the governance and stewardship of the Great Lakes. The International Joint Commission was established in 1909 to advise and negotiate on water resource issues between the United States and Canada. The Great Lakes are an immense natural resource in terms of freshwater, and economic resource through tourism and transportation. The Great Lakes are especially important when considered in the current global water crisis.

Given the abundance of freshwater in the Great Lakes, it may seem out of place to discuss water in the upper Great Lakes in the context of crisis. However, in his book, *Great Lakes Water Wars* (2009), journalist Peter Annin provides a cautionary tale of what *could* happen to the Great Lakes if care is not exercised. In his opening chapter, Annin did not describe a past crisis with the Great Lakes per se but rather focused on the crisis of the Aral Sea and the destruction of that body of water. The Aral Sea was one of the four largest lakes in the world. In the 1960s, a Soviet Union irrigation project diverted water from the rivers that fed the Aral Sea to instead water crops in farms along the rivers, thus greatly diminishing the water supply that replenished the Aral Sea. The Aral Sea has now been reduced to only 10 percent of its original size and has been called "one of the planet's worst environmental disasters" ("Aral Sea …").

The Great Lakes Compact (officially called the Great Lakes-St. Lawrence River Basin Water Resources Compact) was enacted by the U.S. government in 2008. The Compact is an interstate agreement among the states that surround the Great Lakes of the U.S.: Illinois, Indiana, Michigan, Minnesota, New York, Ohio, Pennsylvania, and Wisconsin. The Compact stipulates rules and policies regarding the use of the Great Lakes Basin water. It is significant for this particular chapter as an example of prospective sensemaking in that it was developed in response to threats to the Great Lakes rather than in response to a disaster that already took place.

The Great Lakes Compact was approved by the eight states bordering the Great Lakes in 2007 to 2008. The Compact was introduced to the U.S. Senate on July 23, 2008 by Michigan Senator Carl Levin as S. J. Resolution 45 and was passed through unanimous consent on August 1. It was then passed in the House of Representatives on September 23, 2008, with a vote of 390 in favor, 25 against, and 18 non voting. On October 3, 2008, it was signed into law by President George W. Bush.

While there are several aspects of the Great Lakes Compact discussion that provide a rich resource for analysis, this chapter focuses on the Senate's discussion on September 22 and 23 as that was the only phase where a debate occurred at the federal level with respect to the Compact, and unlike other phases in federal approval, there was not unanimous consent given that 25 representatives voted again the Compact and 18 did not vote. The transcript of the debate can be taken as an instance of prospective sensemaking in that the comments can be viewed as addressing a future anticipated audience and their future anticipated concerns and issues with respect to the Great Lakes.

In addition, meetings are a key activity for organizations and communities (Schwartzman, 1989). They do not just function as events where information is conveyed, but may also be viewed as instantiations of the organization and its hierarchy. Meetings have also been a key area to examine in sensemaking research. For instance, public hearings and inquiries have functioned as vehicles for a community to make sense of what happened during a crisis (Gephart, 2007). Given the significance of meetings as sites for organizational meaning-making (Castor, 2005), organizational conversation and discourse analysts have examined meeting talk as part of understanding the communicative constitution of organizations (Cooren, 2013).

The Great Lakes Compact as Prospective Sensemaking[1]

In this section, I argue that the Great Lakes Compact can be viewed as an example of prospective sensemaking. In addition, I argue that one reason why the Compact was so successful in its passage is because it addressed and tied together several interests and concerns in an "action net" of various account vocabularies surrounding the Great Lakes. The following were identified as key account vocabularies: natural resource, economic development, stewardship, and diversion. I organize the analysis using Stigliani and Ravisi's model of prospective sensemaking by attending to bracketing, articulating, elaborating, and influence.

Bracketing

There has been a long history of concerns surrounding the diversion of water from the Great Lakes (Annin, 2009). For example, the changing of the flow of the Chicago River in 1887 to flow away from rather than into Lake Michigan is actually a water diversion. Annin also documented several other unsuccessful attempts to divert water from the Great Lakes as well as several unsuccessful attempts to enact regulations to prohibit water diversions. Given this long history of concern surrounding the protection of the Great Lakes, what was different about the circumstances of the passage of the Great Lakes Compact in 2008?

In 1998, a private Canadian company had made plans to export 160 million gallons of water from Lake Superior to Asia. This event prompted immediate concern and protest. In the intervening decade from the attempted diversion to the federal Compact approval, an interstate agreement was worked out among the eight U.S. Great Lakes states and approved within each of those states' legislatures and passed by their governors. What is key to note is that the Canadian company's attempt to export water served to "bracket" a situation that could threaten the Great Lakes. The importance of this incident is also evident based on the fact that during the congressional discussion of the Compact, one representative, Mr. Vernon Ehlers of Michigan, recounted this sequence of events:

> The catalyst for the creation of a Great Lakes Compact came in 1998 when the government of Ontario granted a permit to a private Canadian company to ship up to 160 million gallons of water per year to Asia.

Articulating

The aforementioned water shipping event served to help politicians to bracket out a threat to the Great Lakes, however, this still leaves unexplained why at that particular point in time, this event prompted action. The next phase of the prospective sensemaking model, articulating, helps to explain part of this movement from event to potential threat to the Great Lakes.

One key aspect in "articulating" an event is how that event is *framed.* Framing is a linguistic activity where a label is assigned and in that process, certain features of a situation are emphasized. The shipping company's attempt was specifically labeled a "diversion." However, within the context of the Compact, diversions are not wholly prohibited as there are exceptions that would be allowed. Rather, based on the discourse of representatives, there are certain types of diversions that are not allowable. One of the House co-sponsors for the Compact, Ms. Betty Sutton (Ohio) described the type of diversions that the Compact would prevent as "uncontrolled and careless diversions" and "destructive diversions." To understand how a diversion would constitute a threat to the Great Lakes, and thus a potential crisis, it is important to look at the other connections that were made in describing the Great Lakes.

Elaborating

In providing an "elaboration" as part of prospective sensemaking, connections are made in how events are interpreted. One aspect of the Great Lakes that were elaborated on were their importance as a source of freshwater. In his written comments, Representative Ehlers cites the following regarding the Great Lakes:

> The Great Lakes comprise the largest source of freshwater in the world—20 percent of the Earth's total and 95 percent of the surface freshwater in the United States—and they provide drinking water, transportation and recreation to tens of millions of people in the United States and Canada. Although the Great Lakes contain copious amounts of fresh water, less than one percent of the water in the Great Lakes is renewed every year through rain, snow melt, and groundwater recharge, with the remaining ninety-nine percent remaining in the lakes each year. In other words, the Great Lakes are a non-renewable resource that is currently at jeopardy from large-scale water diversions outside the Great Lakes Basin.

There are a number of ways that Mr. Ehlers provided an elaboration and made connections. First, in citing various statistics, Mr. Ehlers highlighted the Great Lakes as a unique natural resource. Second, preservation of the Great Lakes is not just an environmental issue, but it is also an economic issue as Mr. Ehlers highlighted how many people rely on the Great Lakes as a transportation and recreation resource.

Similarly, in her remarks, Ms. Sutton also emphasized the economic importance of the Great Lakes:

> They are also essential for recreation, jobs and the overall health of our economy. Lake Erie alone supports 240,000 jobs and $5.8 billion in wages. ... In addition, the Great Lakes support a multi-billion dollar a year sport fishing and recreational boating industries, and also support travel and tourism throughout the region.

In viewing elaboration from a CCO perspective, another way that elaboration could be described is in terms of associations, a concept put forward in Latour's (2005) work in Actor-Network Theory. The Great Lakes are associated with the environment *and* with economics *and* with recreation *and* with transportation. In addition, the Great Lakes are associated with individuals on a personal level. For example, in his testimony in front of the U.S. Senate Judiciary Committee, Cameron Davis, President and CEO of the Alliance for the Great Lakes and co-chair of the Healing Our Waters®–Great Lakes Coalition stated:

> As a boy, I stood on the shores of Lake Michigan during our Sunday family picnics and marveled at how it seemed like heaven: the blue waters mirrored the sky and the pure white sailboats floated weightlessly, reminding me of angels.

Davis' recounting of his childhood experiences are one of several instances in which I've heard or read of someone describing the Great Lakes in terms individual, personal family experiences.

What I wish to emphasize in this discussion of elaboration is that elaboration occurs through associations. The Great Lakes are significant for their associations as a water resource, transportation facilitator, recreational arena, sport fishing resource, and tourist destination. In other words, in elaboration, there was more than one reason to approve of the Compact and that was because there was more than one association being made.

To borrow from Czarniawska's (2009) discussion of action nets, what is also crucial in all of these associations is that they do not just link agents, actors, or entities. Rather, they are about some type of *activity* or action being performed that the Great Lakes enables.

Influencing

The last stage of Stigliani and Ravisi's model is influencing. In the case of the Great Lakes Compact, this occurred in a number of areas. At the state level, many political representatives and special interest groups mobilized in developing state level Compact agreements. The Great Lake states also worked together to reach a consensus on the Great Lakes Compact. When the Great Lakes Compact finally arrived at Congress a full decade after the precipitating event, it already had the support of the states involved. Also, the Great Lakes Compact, while a U.S. document, included the involvement of the two Great Lakes–St. Lawrence River Canadian provinces of Ontario and Quebec. Perhaps though what is significant to note is where influence occurred in developing the Compact as a legally-binding document in the political realm by giving it the status of federal law.

Summary

This chapter was intended to explicate prospective sensemaking and to describe how this process occurs as a series of communication activities. In line with Gioia et al.'s (2002) work, this chapter illustrated the interconnection of the past and present in sensemaking. Past events functioned as triggers and resources for making sense of possible futures.

Applying Stigliani and Ravasi's (2012) model, the example analysis of the approval of the Great Lakes Compact showed a relationship between prospective and retrospective sensemaking in that an event from the past (an attempted diversion of Lake Superior water) was connected with sensemaking of a potential future (destructive diversions). In the approval of the Great Lakes Compact, there were a multitude of diverse reasons provided to approve the Compact such as the relative scarcity of freshwater globally, the significance of the Great Lakes as a regional source of freshwater, economic reasons such as tourism, sports fishing, and recreation for preserving the Great Lakes. The Great Lakes, in other words, were explained as important in how they were being associated with several activities.

Conclusion

This chapter was intended to examine discourse related to the Great Lakes Compact as an instance of prospective sensemaking. While the Great Lakes Compact is now in place, there are ongoing issues with protection. For example, there were loopholes that were in place in the Compact to allow for certain diversions of water out of the Great Lakes basin. The global water crisis is ongoing and emphasizes the importance of protecting freshwater resources. Even though the Great Lakes Compact was passed, there are still ongoing threats to the Great Lakes ecosystem such as invasive species. While the Great Lakes Compact was intended to protect the Great Lakes, there are ongoing issues that threaten the Great Lakes. Therefore, it is important to continually attempt to project ahead and prospectively make sense of the consequences of various circumstances, actions, or lack of action.

In highlighting prospective sensemaking, I emphasized the importance of projecting into and anticipating the future. Earlier chapters in this work showed how crises can be constructed by missteps in failing to adequately anticipate the future. Rather than focus on failure, this chapter described prospective sensemaking as one model to aid in looking to the future.

The four stages of prospective sensemaking can be re-framed as practical advice. In bracketing, one must be attentive to prior circumstance to understand what is unusual, what can be treated as a warning to further study, or what can be learned from past problems to prevent future problems. In the case of the Great Lakes Compact, this type of bracketing occurred in the attention stage and alarm that was raised through the potential shipping of water. At a different level, this type of bracketing occurred through an awareness of past problems that were created in other areas through diversions. In essence, the practical advice here is to read history, pay attention to the stories of those with more experience, and, be willing to see connections or similarities across different contexts.

In "articulating," one must attend to how they label or frame an issue. For example, as noted in chapter one's discussion of the Mann Gulch fire, it is believed that the fact that the fire was labeled a "10 o'clock fire" led the smokejumper crew to believe that the fire was going to be easily managed. In chapter four, the Georgia leadership resisted framing the winter storm until it was too late to take action to prevent the crisis that ensued.

In "elaborating," it is important to consider what kind of associations or connections can be made across different activities. In other words, how is one activity associated with another? Earlier in the chapter, the example of the Aral Sea was described. In diverting water from the rivers that fed the Aral Sea, one activity (crop irrigation of farmlands along the rivers) was not seen as associated with another activity (the replenishment of the Aral Sea). "Seeing" such associations calls in part for an awareness of the properties of material features, an important interest in Cooren's (2010, 2015) CCO work on relational ontology, materiality, and agency. Cooren's work reminds us that while organizations may be constituted through communication, humans are not the only ones who are involved with constitutive processes. In addition, we must be attuned to how nonhuman agents, whether it is an interesting sculpture as Cooren (2015) described, a weather system, or interconnected lakes, participate in creating circumstances and, potentially, crises.

Finally, in influencing, one must consider how best to have an impact to affect change. The interest in this work is in organizational communication, whether it be in business, nongovernmental agency, or governance settings. This can also include more ambiguously delineated ways of organizing such as through social movements. Ultimately, this calls for being aware of who or what to associate with that can affect future actions to forestall or possibly even prevent a crisis.

Note

1. This analysis is based in part on analysis presented in Castor (2016), a project that was supported by a C-SPAN Education Foundation/Purdue C-SPAN Video Archive Individual Grant, 2015.

References

Annin, P. (2009). *The Great Lakes water wars.* Washington, DC: Island Press.

Aral Sea 'One of the Planet's Worst Environmental Disasters'. (2010, April 5). *Telegraph.* Retrieved from https://web.archive.org/web/20100408214552/http://www.telegraph.co.uk/earth/earthnews/7554679/Aral-Sea-one-of-the-planets-worst-environmental-disasters.html

Castor, T. R. (2005). Constructing social reality in organizational decision making: Account vocabularies in a diversity discussion. *Management Communication Quarterly, 18*(4), 479–508.

Castor, T. R. (2016). Discursively constructing the Great Lakes freshwater. In R. Browning (Ed.), *Advances in research using the C-SPAN archives* (pp. 33–57). West Lafayette, IN: Purdue University.

Cooren, F. (2004). The communicative achievement of collective minding: Analysis of board meeting excerpts. *Management Communication Quarterly, 17*(4), 517–551.

Cooren, F. (2010). *Action and agency in dialogue: Passion, incarnation and ventriloquism.* Amsterdam: John Benjamins.

Cooren, F. (Ed.). (2013). *Interacting and organizing: Analyses of a management meeting.* Mahwah, NJ: Lawrence Erlbaum Associates.

Cooren, F. (2015). In medias res: Communication, existence, and materiality. *Communication Research and Practice, 1*(4), 1–15. https://doi.org/10.1080/22041451.2015.1110075

Czarniawska, B. (2009). Introduction: Action nets. In B. Czarniawska (Ed.), *Organizing in the face of risk and threat* (pp. 1–8). Glasgow: Edward Elgar.

Geist, P., & Chandler, T. (1984). Account analysis of influence in group decision-making. *Communication Monographs, 51*, 67–78.

Gephart, R. P. (2007). Crisis sensemaking and the public inquiry. In C. M. Pearson, C. Roux-Dufort, & J. A. Clair (Eds.), *International handbook of organizational crisis management* (pp. 123–160). Thousand Oaks, CA: Sage.

Gioia, D. A., & Chittipeddi, K. (1991). Sensemaking and sensegiving in strategic change initiation. *Strategic Management Journal, 12*(6), 433–448.

Gioia, D. A., Corley, K. G., & Fabbri, T. (2002). Revising the past (while thinking in the future perfect tense). *Journal of Organizational Change Management, 15*(6), 622–634. doi: 10.1108/09534810210449532

Hill, R. C, & Levenhagen, M. (1995). Metaphors and mental models: Sensemaking and sensegiving in innovative and entrepreneurial activities. *Journal of Management, 21*, 1057–1074.

Latour, B. (2005). *Reassembling the social: An introduction to actor-network-theory.* New York, NY: Oxford University Press.

Marsen, S. (2014). 'Lock the doors': Toward a narrative – semiotic approach to organizational crisis. *Journal of Business & Technical Communication, 28*(3), 301–326.

Putnam, L. L., & Nicotera, A. M. (Eds.). (2008). *Building theories of organization: The constitutive role of communication.* New York, NY: Routledge.

Schoeneborn, D., Blaschke, S., Cooren, F., McPhee, R. D., Seidl, D., & Taylor, J. R. (2014). The threesSchools of CCO thinking: Interactive dialogue and systematic comparison. *Management Communication Quarterly*, *28*(2), 285–316. doi: 10.1177/0893318914527000

Schwartzman, H. B. (1989). *The meeting: Gatherings in organizations and communities*. New York, NY: Plenum Press.

Stigliani, I., & Ravasi, D. (2012). Organizing thoughts and connecting brains: Material practices and the transition from individual to group-level prospective sensemaking. *Academy of Management Journal*, *55*(5), 1232–1259. doi:10.5465/amj.2010.0890

Weick, K. E. (1979). *The social psychology of organizing* (2nd ed.). New York, NY: McGraw Hill.

Weick, K. E. (1995). *Sensemaking in organizations*. Thousand Oaks, CA: Sage.

Wright, A. (2005). The role of scenarios as prospective sensemaking devices. *Management Decision*, *43*, 86–101. doi: 10.1108/00251740510572506

PART 3

CONCLUSION

· 7 ·

CONCLUSION AND PRACTICAL LESSONS

The purpose of this work was to provide an introduction to and illustratrations of how to apply a CCO perspective. In addition, this work focused on climate and weather emergencies to fulfil a number of goals. First, this approach allowed for a discussion of how to re-frame organizational communication-related concepts such as sensemaking, crisis, decision-making, and problem construction in terms of a communication-as-constitutive perspective. Second, it highlighted how climate and weather-related crises are interconnected with issues of organizational communication. At a basic level, these are connected in that, as a matter of public health, organizations, particularly governmental agencies, are at the forefront of identifying, warning, and taking action in the face of climate emergencies. This concluding chapter summarizes key themes related to communication, climate issues, and the CCO perspective. The chapter concludes by addressing criticisms, future directions, and practical lessons based on the examples and crises examined throughout the chapters

Climate Concerns

In chapter two, I explored key influencing theorists on the three main CCO approaches, focusing on the influence of Bruno Latour on the Montreal

School, Anthony Giddens on the Four Flows approach, and Niklas Luhmann and the Social Systems approach. In addition to their important influence on CCO, these scholars have in common a concern with the environment, risk, and social-technological interactions with the aforementioned areas (e.g., Giddens, 2011; Latour, 2004a; Luhmann, 1989).

Climate change is a politically controversial topic (Hoffman, 2015). The debate has some nuances in that some politicians claim that what they dispute are the causes rather than the existence of climate change, however, for others, even the existence of climate change is controversial. Those debates however, do not alter key points in this work in that no matter one's view of the cause, humans still have the capacity to anticipate and prepare for a variety of contingencies and weather and environmental changes. Weather and environmental disasters are not dependent on the existence of climate change. As Beck (1992, 2009) noted, human changes in the environment and technological systems have created our own conditions of risk and vulnerability in circumstances of our own making.

Several key scholars who have influenced the CCO perspective have explicitly addressed concerns regarding climate change. For example, Latour's (2004a) major work related to the environment is titled, *The Politics of Nature: How to Bring the Sciences into Democracy* (also see Latour, 2004b, 2008). Latour addressed the interconnection between science, nature, and politics and argued *against* the "bifurcation" of nature and of the conceptual and philosophical separation of the social and natural spheres. Latour (2008) emphasized the difference between "matters of fact" and "matters of concern" and how we must attend to the latter to understand how matters of concern are "built up" or socially constructed.

Giddens (2011) has written on climate change in a highly acclaimed book, *The Politics of Climate Change*. In this work, he critiques the ability of our current political systems to handle climate change and describes the Giddens's paradox:

> … since the dangers posed by global warming aren't tangible, immediate or visible in the course of day-to-day life, many will sit on their hands and do nothing of a concrete nature about them. Yet waiting until such dangers become visible and acute—in the shape of catastrophes that are irrefutably the result of climate change—before being stirred to serious action will be too late. (p. 2)

The analyses presented in this work serve to illustrate Giddens' points above about how climate change catastrophes are due only in part to climate change, but are also the result of human inaction.

Given Luhmann's interest in social systems (Luhmann, 1984) and risk (Luhmann, 1993), it is understandable that he would apply his thinking to the issue of ecological problems in his work aptly titled, *Ecological Communication* (Luhmann, 1989). Luhmann's interest with environmental problems is one of "how society comes to the very awareness of environmental dangers as such" (Bednarz, 1989, p. xiii). In this respect, Luhmann's interest can be characterized as one of constitution or of how environmental problems are communicatively constituted.

The aforementioned work highlights the interconnection between social, communication, and natural systems. In this project, I have focused on the ways that organizational communication is constitutive and in turn, how these constitutive processes have contributed to the communicative constitution of crises.

The Communicative Constitution of Crises

Examining how crises are communicatively constituted entails analyzing how communication contributes to the shaping and development of a crisis. For example:

- As the case of heatwaves in Europe illustrates, the defining and identifying of crises are complex in that definitions may not be based solely on objective conditions but rather are connected with human impact. In addition, perceptions of crisis may be shaped by timing in that slower moving crises may be more challenging to detect than sudden jolts to a system.
- Timing is connected to crisis response in that to respond effectively, the crisis must be anticipated in order to put appropriate response processes into play. During winter storm Leon, the failure to label and describe the storm as a potential crisis created a situation in which action was taken too late to prevent a crisis: preemptive actions such as calling for a closing of schools and salting of the roads prior to the morning commute did not occur on a timely basis.
- During the Hurricane Katrina disaster, there was confusion over who could act as an agent. This case illustrated how in the face of a crisis, who can take action is something that is communicatively constituted. Elsewhere, I and colleague, Dr. Mariaelena Bartesaghi, identified how metacommunication during emergency manager meetings led to an

overly restrictive meeting format which in turn limited how unanticipated problems could be addressed (Castor & Bartesaghi, 2016).

- Although looking ahead to predict crises is challenging, such an action can be done as the case of the approval of the Great Lakes Compact illustrates. That case illustrated how past cues were taken into consideration regarding the possibility of a crisis. The response was something that was successfully undertaken by creating and calling upon various action nets as part of the persuasive argument for the need of the Compact.

The above examples illustrate a number of ways in which crises are communicatively constituted in problem formulation, the construction of agency, and in social actions. However, a key point I wish to emphasize is that crises are not exclusively *communicatively* constituted. Certain *material* circumstances must also be present for crises to be constructed. Without a hurricane, there is not a hurricane disaster. Without a storm, there is not a winter storm emergency. Without heat, there cannot be a heat wave, and so on. Hurricanes, storms, and extreme heat on their own are not enough to cause a crisis. Rather, the development of a crisis is interconnected with human social systems and their interaction with and responses to those events.

The Communicative Constitution of Organizations

In addition to analyzing climate and weather crises from a communication perspective, this work was intended to more generally illustrate applications of a Communicative Constitution of Organizations or CCO perspective. In an early chapter, I described the three main CCO perspectives of the Montreal School approach, the Four Flows approach, and Luhmann's Social Systems approach. The application chapters drew most heavily from the Montreal School approach given the focus of the application chapters on discourse and language-use during actual meetings. Work within the Montreal School itself is quite diverse with differences among scholars in this area.

The application chapters in this work drew upon and were intended to illustrate a CCO perspective in the following ways:

- Attention to the role of discourse as language-use in interaction in communicatively constituting organizational activities through the analysis of organizational meetings, documents, teleconferences, and email exchanges

- Explication of agency as a capacity of humans and nonhumans by examining the impact that humans and nonhumans have in both shaping crises and in shaping crisis responses
- Analysis of the roles of texts, conversations, and metaconversations (metacommunication) in the communicative constitution of organizations

Criticisms of CCO

The CCO perspective is important in understanding the fundamental role of communication in constituting organizations. In addition, the fact that it is an umbrella concept with room for the development of specific theories to explain CCO is a strength. However, there are of course limitations or challenges generally for CCO perspectives to address. These criticisms can be summarized as follows: demonstrating the "sufficiency" of communication for organization, explaining organizational failures, and materiality.

One challenge in addressing CCO criticisms is that because there are different CCO theories, there are different critiques that have been applied to each. Bisel (2009) has been one of the most vocal and constructive critics of CCO approaches, noting that there are important distinctions between the CCO approaches. Bisel (2010) also noted the important distinction between communication as a "necessary" versus "sufficient" condition for the communicative constitution of organizations. According to Bisel, CCO scholars have argued that communication is necessary for organizations, but they have not adequately addressed sufficiency by identifying the factors that explain how *organization* as opposed to some other social unit is constituted through communication. In other words, CCO scholars have not articulated what is specifically organizational about the communication activities that they have studied. For example, conversations occur all the time in all sorts of contexts to constitute relationships, cultures, families, etc., so what is it about conversations that are specific to the process of organizing?

Another criticism is that CCO scholars have not adequately addressed how communication may prevent an organization from effectively organizing (Bisel, 2010). This is a topic that has been taken up in recent CCO work, most obviously in Taylor and Van Every's (2014) book, *When Organizations Fail: Why Authority Matters*. This work focuses on the idea of authority as a key component in the success or failure of organizations. The role of authority in organizing was also the key theme in a 2016 International Communication

Association pre-conference in Fukuoka, Japan, that took place as a collaboration across the Organizational Communication and Language and Social Interaction divisions.

A third critique is that CCO scholars inadequately have taken into consideration the role of materiality in organizing (see Bisel, 2010; Reed, 2010). However, materialism is another topic that has been taken up by Montreal CCO scholars with a great deal of focus. In particular, Cooren (2015) has addressed this topic as a significant component in his arguments for a relational ontology. Cooren argued that materiality is an important aspect of constitutive processes. Using an example from an art museum, he analyzed the significance of specific statues and artwork in how the meaning of a particular space is communicatively constituted. In this process, Cooren noted that meaning is not solely constituted by human communication; rather, the statues and artwork, their positioning, their composition, etc. also participate in constituting meaning. In this respect, Cooren has identified how materiality is a part of CCO.

Future Directions

Given that CCO encompasses several research perspectives, there are many different directions of future research that are being pursued within this area. There are four main areas of development in this area that I wish to highlight: authority, creativity, technology, and space/place.

Several scholars have focused on the concept of authority as key to understanding how organizations are constituted. Taylor and Van Every (2014) focused on authority as key to understanding why organizations may succeed or fail. Borrowing from other scholars, they identified that there are different forms of authority ("legitimate" and "expert"). In addition, applying ideas from Charles Peirce (1955), they put forward an explanation of authority that described this concept in terms of "orientation to a third party." When organizational members look to a common external party for guidance, they have a common orientation to a "third." This third could be a person, a policy, or something else. When there is a disagreement over a third, there is a failure of authority and a danger to the stability of the organization. Taylor and Van Every studied the failure of a media organization.

The notion of authority is also a key idea taken up in Cooren's (2010) book, *Action and Agency in Dialogue* where he describes his ventriloqual perspective regarding how one may speak on behalf of another. In a related project, Benoit-Barnat and Cooren (2009) addressed authority in terms of "presentification" or the "process of making sources of authority present in interactions" (p. 5). Presentification occurs when a physically distant other is invoked to justify an action. To use an example relevant for students, imagine two undergraduate students are conversing about an assignment that is due for a class and trying to figure out how to complete the assignment. One student says something to the effect of "in class the other day, the instructor said that we needed to find five academic sources." In referring to the instructor, that student is "making present" the authority of the instructor.

Cooren and associates (see Bartels & Bencherki, 2013; Martine & Cooren, 2016; Martine, Cooren, & Bartels, 2015) have explored the idea of creativity to understand its interconnection with communication. When considering issues such as organizational innovation and the creative economy, creativity is a central issue that organizations need to address as part of their ongoing development. Creativity, however, has often been conceptualized as an individual matter dealing with individual psychology rather than as a communication construct. In Montreal School work on creativity, Cooren and colleagues have examined the inter-relationship between creativity, relationality, materiality, and communication. Their work is innovative in that far from leaving creativity to the confines of the individual mind, they have examined how creativity is intertwined with communication.

One distinguishing characteristic of the Montreal School approach is its inclusion of nonhumans in conceptualizing constitutive processes. However, the nonhuman is not just limited to texts, technology, and objects, but also includes other concepts such as space. The theme of space and its connection to CCO work has been explored systematically through the work of Consuelo Vásquez (e.g., Vásquez, 2013, 2016; Vásquez & Cooren, 2013). In one respect, time and space have been a continuous interest in CCO research as scholars have grappled with the issue of organizational maintenance. As Putnam and McPhee (2009) stated, "an important agenda for future research is conceptualizing the role of space in CCO theories" (p. 200). However, this treatment has addressed space in terms of distance between interactions rather than focus on the properties of place and the influence on place and space on organizational interactions.

Vásquez (2013) noted that space has typically been defined in relation and opposition to time where space is equated with absence (e.g., absence of time, movement, change). Alternatively, Vásquez argued for the re-conceptualization of space as performative and relational through the idea of "space-time." In other words, space is constituted through interactions. The implication of this for organizations is that "the organization can transcend, dislocate, displace itself, or tele-act (i.e., act from a distance). It can, in other words, be here and there at the same time" (Vásquez, 2013, p. 128). It is through communication that organizational members create "bridges" across distance in organizational activities. As an example, Vásquez examined the Chile space program and how, in interactions, organizational members attempt to project a science education program across time and space. In her example, there was a problem of a lack of participants for the program to achieve its goals. Through communication and texts such as invitations and forms, advocates of the program attempted to connect and expand their program across Chile. Vásquez demonstrated how space is not a given, but a matter of communication.

Practical Lessons

While the CCO perspective has pushed the development of theorizing about organizations, there are also some important practical lessons that students and practitioners in organizations can take away generally from CCO and specifically from themes discerned from the chapters and cases presented in this work. In many ways, the themes are oversimplifications, however, the themes are presented recognizing that they are partial guides and not absolute formulas for success. These lessons for practice are summarized as: attention to language-use and evidence, organizations are social constructions, study history, anticipating the future, networking, and improvisation.

Attention to Language-Use and Evidence: Language *and* "Matter" Matter

Among the CCO approaches, the Montreal School has an especially strong emphasis on language-use in interactions. Language shapes perception and action. The case studies of this work illustrate the implications of this for organizational actions, especially in light of crises and extreme weather emergencies. Language helps in defining whether or not something is a problem and who or what can take action to respond to and solve or even preempt a problem.

However, language alone is not responsible for constituting reality as there are material circumstances and physical entities such as weather events and technology systems that participate in constitutive processes. Therefore, language matters for both shaping what counts as a problem and for what may not be considered a problem, but should be. Language both enables *and* constrains constitutive processes. Subsequently, just because something isn't identified or defined as a problem at the moment, does not mean it will not evolve into a problem in the future.

In addition, "matter" matters. Physical, material entities and circumstances function to shape and constitute organizational reality. To illustrate this, I refer to an example from a work of fiction that, nonetheless, illustrates this point. In the film, *Manderlay*, written and directed by Lars von Trier, the main character, Grace Mulligan (played by Bryce Dallas Howard) arrives at a post-Civil War plantation in which there is still slavery. The movie is intended to be filmed as an allegorical rather than "realistic" story. Grace is the daughter of a gangster who takes it upon herself to prepare the former slaves for life off of the plantation as free people. However, a number of things go wrong in this supposedly well-intentioned plan. The specific scene of relevance here is one in which there is a disagreement on the plantation regarding what time it is. Grace decides to have a vote, and the plantation residents happen to vote on a time that is ten minutes off of the "real" time. Subsequently, when Grace's father arrives back to pick her up, she is not ready for him.

In one sense, you could say that the plantation residents "socially constructed" time through their communication practices (a vote). However, this specific, localized social construction did not mesh with other external agreements on defining the time. Similarly with crises, humans may attempt to define situations in particular ways, however, those social constructions may not mesh with material circumstances.

Organizations Are Socially Constructed

While the CCO perspective emphasizes organizing as a collective or group, just because a collective is put forward does not mean that people are limited to the level of staying at the collective. In other words, while individual actions contribute to the formation of organizations there are opportunities to challenge the organization and/or particular actions taken on behalf of the organization. To further explain this, I turn back to the opening examples and quotations from chapter one.

The statement, "it was an organizational decision," refers to the organization as a collective that was responsible for a decision. In this respect, this statement can be seen as contributing to the communicative constitution of the organization and the attribution of agency to the organization. However, this was not satisfactory to Green Bay Packer football fans who wanted more information such as who exactly in the organization was responsible for the decision and why did he make that particular decision. The situation was not helped through the nonverbal vocal inflections and overly neutral facial expressions of Coach Mike McCarthy during the press conference as he gave the impression of an overly rehearsed speech when he referred repeatedly to the "organizational decision." Eventually it came out that the lead decision-maker was the Green Bay Packer's general manager, Ted Thompson.

My point in discussing this example is not to be prescriptive in directing attention to the role of individuals in constituting organizations. The key idea is that how organizations, decisions, problems, crises, etc. are communicatively constituted and can be questioned at any point. This is also a key point that my co-researcher, Mariaelena Bartesaghi, and I made in a chapter on social constructionism (see Bartesaghi & Castor, 2008). One of the characteristics of a communication social construction perspective that we identified is that it is in *questioning* that we can use our communication practices to challenge how and why something is constructed or constituted in particular ways. This was also an important point made by Latour (2004b) when he called for an emphasis on *construction* rather than *deconstruction*. The latter idea focuses on taking apart and critique. In focusing on *construction*, Latour calls for understanding how "things" are "put together."

Study History

An important emphasis in the Montreal School approach is how texts, conversations, and metacommunication can bring the "there and then" into the "here and now." In other words, something that happened in the past in a different setting can be brought into the present through communication. This is important as the past can provide many lessons and insights into handling and understanding a current dilemma.

There are many layers to the past though. In considering the communicative constitution of organization, the past consists of the organization's history of past conversations, texts, and other events. However, as McPhee and Zaug's (2000) work in the four flows model highlights, organizations exist within a societal context. Therefore, I also advocate looking to past conversations, texts, and other events that occurred elsewhere that would be relevant for an issue that the organization is addressing.

To illustrate the above points, there are two examples from the Hurricane Katrina disaster that are relevant. There was a great deal of blame cast as to who was responsible for the delay in responding to the disaster. The cause of the disaster is complex with no single person or agency responsible. For this illustration, I focus on the role that New Orleans Mayor Ray Nagin played during the disaster. As Hurricane Katrina was approaching, there was a delay in Mayor Ray Nagin's call and decision to evacuate the city of New Orleans (Harris, 2005). According to a report issued by the Select Bipartisan Committee to Investigate the Preparation for and Response to Hurricane Katrina (2006), "as late as Saturday afternoon, according to news reports, Nagin was consulting city lawyers about legal liability to the city's businesses for lost revenue from evacuating customers" (p. 109). However, according to the New Orleans Comprehensive Emergency Plan, the mayor has the authority to issue an evacuation order. To view Mayor Nagin's delayed decision in terms of a text-conversation perspective and the guideline to "study history," Mayor Nagin demonstrated a lack of awareness or understanding of a text that had been previously created to guide the city mayor's in the exact circumstance that he faced of calling for an evacuation in the face of an emergency.

"Studying history" calls for being attuned to the organizational texts and prior organizational conversations that were developed to or may guide organizational members during decision-making situations. It calls for being aware of organizational history, policies, and procedures.

Anticipating the Future

"Anticipating the future" is difficult to do as it calls for being able to make predictions about the future. While this is challenging, it is not impossible to do. This is an area of organizational activity that organizations may expend a great deal of energy on as organizational members develop strategic plans and forecasting models. Looking to the future is something that is important for organizational members to do to anticipate and plan for future events.

The examples related to weather crises that were featured in this work illustrate the importance of looking ahead to outcomes that are likely to happen and putting into place preparations that anticipate those outcomes. In some situations the outcomes of a disaster can even be prevented if appropriate human actions are taken such that, for example, the coming of a hurricane became a city-wide evacuation rather than large scale human disaster, or the coming of a winter storm becomes a school and business cancellation day rather than a crisis.

Networking

Action nets refer to the connections and associations that are formed between different agents. As some of the case studies have illustrated, effective organizational actions are dependent upon mobilizing an action net. For example, during a weather disaster, typically, a governmental agency would need to know how to work with other types of organizational groups such as hospitals, media groups, weather forecasters, and non-governmental aid organizations. Action nets are crucial for responses to crises. However, action nets are also important to rely upon when planning pre-emptively to *prevent* a disaster or crisis.

There are a number of ways that emergency managers and other crisis responders can effectively mobilize their networks. It is important to build ties or connections ahead of time, before a crisis strikes to understand who or what agency may be called upon to do what and what roles may be assumed. In this way, such a priori networking can help in "practicing for" a crisis ahead of time. Prior to Hurricane Katrina, this is in part what the Hurricane Pam scenario exercises were intended to accomplish in anticipation of a major hurricane until the exercises were discontinued due to a lack of federal support (Cooper & Block, 2006).

Improvising

Improvisation is commonly associated with jazz music when musicians create new music on the spot as they are playing. Improvisation is also an important concept that has been appropriated in organizational communication to refer to how organizational members need to be flexible in order to innovate and adapt to unanticipated circumstances (e.g., Weick, 1993). Improvisation involves creativity and spontaneity, which makes it especially helpful when dealing with new circumstances. This concept is one that Weick (1993) applied in his analysis of the Mann Gulch fire to account for how one smokejumper was able to survive this deadly fire: in the unique circumstances of this deadly fire, the smokejumper improvised and developed a new technique by creating an "escape fire."

In improvising, it is important to note that improvisation does not necessarily mean creating something wholly new and never before created. Improvisation can be based on drawing from an old procedure or technique or pattern, but doing some modifications or changes in how it is applied. In this respect, improvisation is related to creativity and learning (Barrett, 1998). While it has some grounding in the past and in a community of practice, there is enough flexibility involved such that improvisation is not constrained by rigid structures and patterns from the past. Perhaps what is most surprising about improvisation is that it also involves practice (Mirvis, 1998).

In summary, improvisation is important generally for organizational success but specifically when dealing with a crisis that involves new circumstances. Improvisation involves elements from the past in addition to breaking away from the past. It also involves practice that takes place within a community.

Summary of Practical Lessons

In summary, the practical lessons may be described as follows:

1. Language *and* physical/material circumstances *both* construct reality. Therefore, practitioners should be attuned to how they communicate and the ways that words can both open up but also limit options. Practitioners also need to be aware of and understand how physical circumstances (e.g., a storm) will shape and constrain actions.
2. Organizations do not just exist on their own, but are shaped and socially constructed by people and communication. Organizations can be both constructed and deconstructed. The organization is not an entity that takes on a life of its own, even though people may act as it does. Organizations are social collectives that depend on people and communication.
3. It is important to understand historical context at many levels (e.g., of individuals, issues, organizations, policies, political economies, etc.). The past can provide insights into the present as well as an understanding of what actions can or should be taken.
4. It is important to think about the future or what could happen as part of being prepared for the expected *and* the unexpected.
5. Practitioners should understand how their organization relates to and is interconnected with other organizations and issues.
6. Being able to improvise is important for organizational learning especially when faced with a crisis.

Closing Notes

The overall purpose of this work was to provide an introduction to a communication as constitutive approach to understanding organizing and organizations. For this purpose, I provided an overview of the main CCO approaches, but focused on applying concepts and ideas from the Montreal School approach in order to examine the details of talk and interaction when people are "on the ground" to interact with each other. A key focus of this work was on cases and crises that related to extreme weather crises, natural hazards, and the environment. These cases were selected to provide a specific topical theme, and they were also selected because of the pressing importance of these topics. There is still uncertainty regarding changes in the planet's climate. However, there are also tools and actions that can be taken to address extreme weather issues and the environment in order to minimize harm.

References

Barrett, F. J. (1998). Coda—Creativity and improvisation in jazz and organizations: Implications for organizational learning. *Organization Science*, 9(5), 605–622.

Bartels, G., & Bencherki, N. (2013). Actor-network theory and creativity research. In *Encyclopedia of creativity, invention, innovation and entrepreneurship* (pp. 29–35). New York, NY: Springer.

Bartesaghi, M., & Castor, T. R. (2008). Social construction in communication: Revisiting the conversation. *Communication Yearbook*, *32*, 3–39.

Beck, U. (1992). *Risk society: Towards a new modernity* (M. Ritter, Trans.). London: Sage.

Beck, U. (2009). *World at risk* (C. Cronin, Trans.). Malden, MA: Polity Press.

Bednarz, J. (1989). Translator's introduction. In N. Luhmann (Ed.), *Ecological communication* (J. Bednarz, Trans.). Chicago: University of Chicago.

Benoit-Barnat, C., & Cooren, F. (2009). The accomplishment of authority through presentification: How authority is distributed among and negotiated by organizational members. *Management Communication Quarterly*, *23*(1), 5–31.

Bisel, R. S. (2009). On a growing dualism in organizational discourse research. *Management Communication Quarterly*, *22*(4), 614–638. https://doi.org/10.1177/0893318908331100

Bisel, R. S. (2010). A communicative ontology of organization?: A description, history, and critique of CCO theories for organization science. *Management Communication Quarterly*, *24*(1), 124–131. https://doi.org/10.1177/0893318909351582

Castor, T., & Bartesaghi, M. (2016). Metacommunication during disaster response: "Reporting" and the constitution of problems in Hurricane Katrina teleconferences. *Management Communication Quarterly*, *30*(4). https://doi.org/10.1177/0893318916646454

Cooper, C., & Block, R. (2006). *Disaster: Hurricane Katrina and the failure of Homeland Security*. New York, NY: Times Book.

Cooren, F. (2010). *Action and agency in dialogue: Passion, incarnation and ventriloquism*. Amsterdam: John Benjamins.

Cooren, F. (2015). In medias res: Communication, existence, and materiality. *Communication Research and Practice*, *1*(4), 1–15. https://doi.org/10.1080/22041451.2015.1110075

Giddens, A. (2011). *The politics of climate change* (2nd ed.). Cambridge: Polity.

Harris, D. (2005, September 6). Who's to blame for delayed response to Katrina? *ABC News*. Retrieved from http://abcnews.go.com/WNT/HurricaneKatrina/blame-delayed-response-katrina/story?id=1102467

Hoffman, A. (2015). *How culture shapes the climate change debate*. Stanford, CA: Stanford University.

Latour, B. (2004a). *Politics of nature: How to bring the sciences back into democracy* (C. Porter, Trans.). Cambridge, MA: Harvard University.

Latour, B. (2004b). Why has critique run out of steam?: From matters of fact to matters of concern. *Critical Inquiry*, *30*(2), 225–248. https://doi.org/10.1086/421123

Latour, B. (2008). What is the style of matters of concern? *Two lectures in empirical philosophy*. Department of Philosophy of the University of Amsterdam, Amsterdam: Van Gorcum.

Luhmann, N. (1984). *Social systems* (J. Bednarz with D. Baecker, Trans.). Stanford, CA: Stanford University.

Luhmann, N. (1989). *Ecological communication* (J. Bednarz, Trans.). Chicago: University of Chicago.

Luhmann, N. (1993). *Risk: A sociological theory* (R. Barrett, Trans.). New York, NY: Aldine de Gruyter.

Martine, T., & Cooren, F. (2016). A relational approach to materiality and organizing: The case of a creative idea. In L. Introna, D. Kavanagh, S. Kelly, W. Orlikowski, & S. Scott (Eds.), *Beyond interpretivism? New encounters with technology and organization* (pp. 143–166). Cham, Switzerland: Springer International.

Martine, T., Cooren, F., & Bartels, G. (2015). *Assessing creativity through its degrees of objectivity: An actor-network approach to creativity*. Paper presented at the 65th International Communication Association Annual Conference, San Juan, Puerto Rico.

McPhee, R. D., & Zaug, P. (2000). The communicative constitution of organizations: A framework for explanation. *Electronic Journal of Communication*, *10*(1/2), www.cios.org/getfile/McPhee_V10n1200.

Mirvis, P. H. (1998). Variations on a theme—Practice improvisation. *Organization Science*, *9*(5), 586–592.

Peirce, C. S., & Buchler, J. (1955). *Philosophical writings of Peirce*. New York, NY: Dover.

Putnam, L. L., & McPhee, R. (2009). Theory building: Comparisons of CCO orientations. In L. L. Putnam & A. M. Nicotera (Eds.), *Building theories of organization: The constitutive role of communication* (pp. 187–207). New York, NY: Routledge.

Reed, M. (2010). Is communication constitutive of organization? *Management Communication Quarterly*, *24*(1), 151–157.

Select Bipartisan Committee to Investigate the Preparation for and Response to Hurricane Katrina. (2006). *A failure of initiative*. Washington, DC: U.S. Government Printing Office.

Taylor, J. R., & Van Every, E. J. (2014). *When organization fails: Why authority matters*. New York, NY: Routledge.

Vásquez, C. (2013). Spacing organization, or how to be here and there at the same time. In D. Robichaud & F. Cooren (Eds.), *Organization and organizing: Materiality, agency and discourse* (pp. 127–149). New York, NY: Routledge.

Vásquez, C. (2016). A spatial grammar of organising: Studying the communicative constitution of organisational spaces. *Communication Research and Practice*, *2*(3), 351–377. https://doi.org/10.1080/22041451.2016.1221686

Vásquez, C., & Cooren, F. (2013). Spacing practices: The communicative configuration of organizing through space-times. *Communication Theory*, *23*(1), 25–47. https://doi.org/10.1111/comt.12003

Weick, K. E. (1993). The collapse of sensemaking in organizations: The Mann Gulch disaster. *Administrative Science Quarterly*, *38*(4), 628–652.

INDEX

A

B

C

R

S

T

V

W

www.ingramcontent.com/pod-product-compliance
Ingram Content Group UK Ltd.
Pitfield, Milton Keynes, MK11 3LW, UK
UKHW021058130926
8556IPUK00046B/1155

9 781433 133350